Recent Advances In Plant Protection

Authors

Dr. N. Arun Kumar

M.Sc (Ag.), Ph.D., Department of Agricultural Microbiology, Tamil Nadu Agricultural University, Coimbatore - 641003, Tamil Nadu, India

Dr. C. Selvi

M.Sc (Ag.), Ph.D., Senior Research Fellow, ICAR - Central Institute for Cotton Research, Regional station, Coimbatore - 641003, Tamil Nadu, India

Dr. G. Guru Prasanna Pandi

M.Sc (Ag.), Ph.D., Scientist (Division of Crop Protection), ICAR-National Rice Research Institute, Cuttack-753006, Orissa, India

Dr. T. Boopathi

M.Sc (Ag.), Ph.D., Scientist SS (Agricultural Entomology), ICAR - Research Complex for NEH Region, Mizoram Centre, Kolasib –796081, Mizoram, India

AkiNik Publications

New Delhi

Published By: *AkiNik Publications*

AkiNik Publications
169, C-11, Sector - 3,
Rohini, Delhi-110085, India
Toll Free (India) – 18001234070

Author: *Dr. N. Arun Kumar, Dr. C. Selvi, Dr. G. Guru Prasanna Pandi and Dr. T. Boopathi*

1st Edition: *2018*

Pages: *171*

ISBN: *978-93-88112-46-8*

Price: *₹515/-*

Preface

In the recent years, increasing demand of food production is attributed to the burgeoning population. Limited land resource stipulates intensive farming system which led to the development of various pest and diseases, thereby making plant protection inevitable. Plant Protection is a multidisciplinary approach that continues to play a significant role in achieving targets of crops production. Integrated Pest Management being the major thrust area of plant protection, ensures availability of safe and quality pesticides to combat the pests and diseases; which otherwise could have a devastating effect on farmer livelihoods. Despite providing quality pesticides, it also streamlines the quarantine measures to eliminate the chances of entry of exotic pests.

Climate change, a global concern particularly the sharp increase in CO2 concentration, has increased the susceptibility of plants to pathogens and pests. Furthermore, short lifecycle and poikilothermy, insects often exhibit rapid responses to changes in weather variables. Very few studies have addressed the changes in the crop-pest interactions that could arise due to the anticipated climate change. Due to chemical intensive agriculture, our ecosystem is heavily contaminated, thus affecting the crops adversely. This has encouraged researchers to look for eco-friendly and novel approaches for pest management and also emphasise the need of R and D in these arena. This book is a ready reference for students, scientists, researchers and extension workers.

The information on recent advances in plant protection is scattered. The book as such is a humble effort to comprehend the latest developments in this direction. This volume has been divided into 9 Chapters focussing on different aspects of plant protection- microbiology, plant pathology and entomology. Within the broad area, invasive species; Plant diseases: Novel Detection and Diagnosis methods; Impact of Climate change on insects; Invasive Insect pest-Spiralling Whitefly; Biological control of brinjal shoot and fruit borer; Microbial biopesticides for insect pest management in cotton; Host Plant Resistance in legumes against bruchids and Role in Soil Nutrition and Plant Protection are elaborated. Relevant data in all chapters have been supplemented with tables and graphical presentations. Efforts have also been made to bring uniformity in the contents of the chapters. The information contained here is a reflection of the compilation of various works done by different researchers across the globe. The editorial board is especially indebted to all the authors who have put their best efforts in gathering relevant information from all the possible sources.

Although we have tried our best to present this information in a perfect manner as possible, and is open to improvement. As editors, we have tried to comprehend the information error free and are awaiting for readers to point out the lacunae if any. We would be happy to address them in the revised edition.

N. Arun Kumar

C. Selvi

G. Guru Prasanna Pandi

T. Boopathi

Table of Contents

Chapter - 1
Climate Change on Insect Pest: An Overview

Authors

Guru Prasanna Pandi G.
Division of Crop Protection, ICAR-National Rice Research Institute, Cuttack, Odisha, India

Soumia P.S.
Division of Crop Protection, ICAR-Directorate of Onion and garlic Research, Pune, Maharashtra, India

Raghu S.
Division of Crop Protection, ICAR-National Rice Research Institute, Cuttack, Odisha, India

Mayabini Jena
Division of Crop Protection, ICAR-National Rice Research Institute, Cuttack, Odisha, India

Sujithra M
Division of Entomology, Indian Agricultural Research Institute, New Delhi, Delhi, India

Subhash Chander
Division of Entomology, Indian Agricultural Research Institute, New Delhi, Delhi, India

Chapter - 1

Climate Change on Insect Pest: An Overview

Guru Prasanna Pandi G, Soumia P.S, Raghu S, Mayabini Jena, Lopamudra Mandal, Sujithra M. and Subhash Chander

1. Introduction

Climate change, a current global concern is the change in climate over time, either due to natural phenomenon or because of anthropogenic activity and exerts multitude of threats to many species in various forms.Very few studies have addressed the changes in the crop-pest interactions that could arise due to the anticipated climate change. Besides, regional impacts of climate change would vary significantly; there is a need for crop-specific and region-specific analysis. Owing to their short lifecycle and poikilothermy, insects often exhibit rapid responses to changes in weather variables and are therefore ideal candidates to model crop-pest interactions under various climate change scenarios.As an adaptation to climate change, changes in package of practices of rice such as sowing time and pesticide application rates may have to be revised to effectively suppress the pest population and sustained agricultural production. Thus calls for meticulous research planning and development of suitable pest management strategies are warranted to ensure food security in future.

2. Changing Climatic Condition: A Global Perspective

The Intergovernmental Panel on Climate Change (IPCC, 2001) reported that increased anthropogenic activities have been responsible for global warming observed over the last five decades. The global mean surface temperature already rose by 0.6±0.2°C during 20th century. It is also projected that temperature rise by about 2°C over the next 100 years, would extend the negative effects of global warming to the most regions of the world. The possible biological consequences of global warming would not only be due to increased temperature but more importantly due to the rate of temperature increases (Root and Schneider, 1993). Besides, increase in atmospheric CO_2 concentrations and the associated temperature rise, and the precipitation patterns will have profound effects on terrestrial plant growth and productivity in near future (Reddy *et al.*, 2010).

The pre-industrial levels of carbon in atmosphere rose from 285ppm (600 giga tonnes (Gt) to the current level of 384ppm (800 Gt), which is further predicted to rise to 570ppm (1000 Gt) by the end of 2050 (IPCC, 2007). Such an abnormal rise in the atmospheric CO_2 levels would result in direct and indirect global climate change. Climate change will also have a profound influence on insect populations along with other living systems. Temperature will affect insects directly through its effect on their growth and development, and indirectly by influencing host plants. The CO_2 would thus affect insects indirectly through host plants by influencing the host physiology.

3. Effect of Climate Change on Plants

Recent reports suggested that the possible effects of doubling of the current atmospheric CO_2 levels could be substantial (Coviella and Trumble, 1999), as elevated atmospheric CO_2 would have 'fertilization effect" on crop plants especially C_3 plants (La Marche *et al.,* 1984). CO_2 has a positive effect on photosynthesis and photosynthates production in various plants species (Grononger *et al.,* 1996). However, the photosynthesis rates increase initially but may decrease after a period of time (Poorter *et al.*, 1988; Dugal *et al.*, 1990; Hall and Allen, 1993). The dark respiration and photorespiration rates also decrease under high CO_2 (El Kohen *et al.*, 1991; Coviella and Trumble, 1999) resulting in change of nitrogen concentration in plants (Lawler *et al.*, 1997) and increase in carbohydrate levels (Osbrink *et al.*, 1987; Ahmed *et al.*, 1993; Mitchell *et al.*, 1993). An accelerated rate of photosynthesis in plants was evident under high CO_2 level that altered the carbon to nitrogen (C: N) ratio. Under such changed C: N ratio, the herbivore insects have been found to have higher feeding rates in order to meet their nitrogen requirements. Soybean plants grown in Free Air CO_2 Enrichment (FACE) facility with high CO_2 produced more carbohydrates and less nitrogen that attracted more Japanese beetles, western corn rootworms and aphids (Srinivasa Rao, 2009).

The increased photosynthetic rates may result in more biomass accumulation, more photosynthetic activity and therefore, plants can potentially compensate for the insect damage (Trumble *et al.*, 1993). A review of 770 experimental observations on numerous crops grown in elevated CO_2 atmosphere documented 14 and 34 per cent average yield increases for C_4 and C_3plants, respectively, (Coviella and Trumble, 1999). Ziska and Teramura (1992) indicated that growth at elevated CO_2 resulted in a significant decrease in night time respiration and increase in photosynthesis, total biomass and rice yield. However, in plants exposed to simultaneous elevated CO_2 and ultraviolet-B (UV-B) radiation, CO_2 enhancement effects on respiration, photosynthesis and biomass has nullified or significantly reduced.

Studies have also shown a significant increase in seed production of pulses due to positive effect of elevated CO_2 on plant development (Miyagi *et al.*, 2007). The effects of elevated CO_2 on physiological processes involving photosynthetic enhancement and stomatal conductance have been demonstrated to affect plant development (Piikki *et al.*, 2007). Reekie and Bazzaz, (1991) reported that *Guara brachycarpa* showed developmental delay by reducing flowering time when exposed to elevated CO_2 concentrations. Rice grain yield was found to decline by 10 per cent for each 1^0C increase in growing-season minimum temperature in the dry season from 1992 to 2003 (Peng *et al.*, 2004). Significant variations in the physiological, biochemical and molecular responsiveness of plants to elevated CO_2 atmosphere were observed. The C_3 plants grown in pots under elevated CO_2 showed photosynthetic acclimation due to soil and nutrients limitation associated with reduced root volume. However, the experiments conducted in open top chambers (OTCs) and FACE showed increase in light saturated rates of photosynthesis in several C_3 plants grown at elevated CO_2 and marked increase in net assimilation rates due to intercellular CO_2 concentrations (Reddy *et al.*, 2010).

4. Impact of Climate Change on Insect Pest and Natural Enemies

In general, two approaches have mainly been used to analyze climate change effects on tri-trophic interactions. One approach is to apply mathematical models to predict the response of specific groups of natural enemies to climate change (Stireman *et al.*, 2005; Guiterrez *et al.*, 2008) and the other approach plans to provide empirical proof linking all three trophic levels *viz.*, plant-insect-natural enemy through experiments to demonstrate the responses to individual climate change parameters, either elevated CO_2 or temperature (Bezemer *et al.*, 1998a). As far as insects are concerned, climate change may impact crop-pest synchrony, interspecific interactions, geographical distribution, population growth rates, period of the activity, and number of generations, overwintering success and invasion by migrant pests (Memmott *et al.*, 2007; Parmesan, 2007).

4.1 Effect of Temperature on Insects

Insects are cold-blooded organisms; temperature of their bodies being approximately the same as that of the environment. Therefore, temperature is probably the single most important environmental factor influencing insect behavior, distribution, development, survival and reproduction (Das *et al.*, 2011). Insect life stage predictions are most often made by using accumulated degree-days (DD) from a base temperature and biofix point (Charles *et al.*, 2006).

Bale *et al.* (2002) opinioned that the effects of other environmental factors on insects could easily overwhelmed by the effect of temperature. Yamamura and Kiritani (1998) predicted that insects might have 1-5 additional life cycles/season with each 2°C increase in temperature. Therefore, climate change resulting in increased temperature could influence crop pests in several complex ways. Although in tropical climates where insects are already living near their higher limit of favourable temperature range, temperature rise might tend to depress insect populations there. However, warmer temperature in temperate climate would result in more types and higher populations of insects (Neumeister, 2010).

Increased temperatures can potentially affect insect survival, development, geographic range and population size (Porter *et al.,* 1991). Temperature can directly affect insect growth and development and indirectly through changes in host plant physiology. It can exert different effects on insect species depending upon their development "strategy" (Bale *et al.,* 2002). Over the course of life span, perennial insects would tend to adjust with moderate temperature fluctuations, while crop pests with "stop and go" development strategy in relation to temperature, would develop more rapidly during periods of suitable temperatures. Increased temperature will accelerate the development of insects possibly resulting in more generations and eventually leading to more crop damage (Diku and Mucaj, 2010). Lewis (1997) reported that there might be reduction in parasitism if thrips populations emerge and pass through vulnerable life stages before emergence of parasitoids and changes in the gender ratio of thrips was also found due to altered temperature. Bale *et al.* (2002) observed that insects spending important stages of their life cycle in the soil might be affected more gradually by temperature changes compared to above ground pests because soil provides an insulating medium that tends to buffer temperature changes more than the air.

Insects might experience lower winter mortality due to increased temperatures thereby considerably increasing the insect population levels (Harrington *et al.* 2001). Higher average temperature might create favourable environment for some crops to grow further north, thus influencing at least some of the insect pests of those crops to shift to the expanded crop areas. Insect species diversity seems to decrease in an area with higher latitude and altitude (Andrew and Hughes, 2005). However, rising temperature could result in more insect species attacking more hosts in temperate climate characterized by higher latitude and altitude. Based on evidence developed through fossil record study, the diversity of insect species and the intensity of their feeding had increased historically with increasing temperature (Bale *et al.,* 2002).

Development, oviposition and longevity of whiteflies were closely related to temperature and there was more delay in development from eggs to adults at low temperatures (Butler *et al.*, 1983; Wang and Tsai, 1996). Some insects are closely associated to a specific set of host crops. Temperature increase may force farmers not to grow certain host crop anymore, which can result in decreased populations of insect pests specific to those crops. Besides, the factors that influence insect pests can also positively affect their natural enemies as well as the disease organisms that infect the pests, thereby resulting in increased attack on insect populations. At higher temperatures, greater predation of aphids occurs, as they are less responsive to alarm pheromone when attacked by natural enemies (Awmack *et al.*, 1997).

A prediction states that increase in the mean surface temperature by 1.0° C during the last 40 years in Japan has resulted in the upsurge of rice and fruit infesting bugs and their pole ward geographic spread. It has been reported that *Nezara viridula* and *Halyomorpha halys* adults experienced 15 per cent reduction in winter mortality with every 1°C rise. Northward expansion of over 50 butterfly species has been recorded with previously migrant 10 species having been established on Nansei Islands between the periods of 1966–1987. It has also been reported that due to global warming there has been a decline in *Plutella xylostella* abundance and increase in natural enemies of *Helicoverpa armigera*and *Trichoplusia ni* barring spiders due to increase in their number of generations (Kiritani, 2006).

Temperature is the sole important environmental factor that determines the insect growth, development, survival, reproduction, distribution and behaviour. In general, temperature impact on insects largely engulfs other environmental factor impacts (Bale *et al.*, 2002). It has been predicted that with the temperature increase of 20C insects will undergo one to five additional life cycles/season (Yamamura and Kiritani, 1998). In the context of climate change, temperature affects insects directly and CO_2 affects them indirectly through host plants (Netherer and Schopf, 2010). Increased atmospheric CO_2 has profound impact on C_3 plants such as rice primarily due to positive changes in photosynthetic carbon assimilation pattern. The enhanced photosynthetic activity leads to increased biomass and ultimately increased productivity (Long *et al.*, 2006; Hasegawa *et al.*, 2007; Zhang *et al.*, 2013). Along with plant quality, herbivore performance is also affected by changes in atmospheric CO_2 (Awmack *et al.*, 2004). Under elevated CO_2, carbon amount in plant tissues is relatively more with corresponding increase in the C: N ratio that causes reduction in plant protein concentration thereby lowering the availability of nutrients to the herbivores (Lincoln *et al.*, 1986; Zhang *et al.*, 2013). To compensate for the inferior nutritional quality in plants, the herbivores increases their food consumption (Fajer, 1989; Marks and Lincoln, 1996; Srinivasa Rao *et al.*, 2009; Johnson *et al.*, 2014). Besides, size of plant canopy may also influence pest populations especially those of sucking pest such as rice planthoppers through changes in micro-environment. However, actual effect on plant growth and yield would depend upon the interaction between CO_2 and temperature.

4.2 Impact of Elevated CO_2 on Insect Pests

Global warming will affect growth and development of all organisms including insect pests. Climate change may disrupt not only pest dynamics in agriculture, but also the dynamics of herbivores in stable ecosystems. Netherer and Schopf (2010) observed that ectothermic organisms have been greatly influenced by changing climate condition directly with respect to development, reproduction, dispersal and mortality and indirectly through changes in plant nutrition, resistance and community interactions. Climate change might result in alteration in geographical distribution, increased overwintering, reduction in generation time, changes in population growth rates, increase in number of generations, extension of the development season, changes in crop-pest synchrony, changes in inter-specific interactions and increased risk of invasion by migrant pests.

Understanding of alteration in insect feeding behavior due to elevated CO_2 and O_3 will be important for predicting crop productivity as well as for identifying insect species likely to become pests in the future (Baker *et al.*, 2000). Elevated CO_2 would have an indirect effect on insect communities as changing host plant phenology may create greater imbalance in insect-plant and tri-tropic interactions. Both warmer weather and elevated CO_2 possibly modify the bionomics of insects under the changing climate (Hulle *et al.*, 2008). An increased atmospheric CO_2 concentration affected growth and development of chewing insects that resulted in increased developmental time, reduced fecundity and decreased larval and pupal weights (Watt *et al.*, 1995). Whittaker (1999) observed that leaf chewers were generally able to compensate for lower quality of food by increasing consumption up to 25 per cent without adverse effect on pupal weight, while sap feeders were the only functional group to show positive response to elevated CO_2.

Projected increase in global mean temperature and atmospheric CO_2 likely to manipulate insect-plant interaction. Elevated CO_2 and elevated temperature have directly affected plant traits such as nitrogen content that is crucial to insect herbivores, whereas temperature alone affected insects. Flynn *et al.* (2006) stated that insect populations did not change significantly under elevated CO_2, but tended to increase only slightly. Plant growth rates were enhanced before infestation, while the height and biomass were not significantly affected by the CO_2 treatment. These results indicate that the combined effects of both elevated temperature and CO_2 may aggravate pest damage to plants and particularly in the plants that responsed weakly to the rising atmospheric CO_2. Enhanced CO_2 level triggered biochemical and physiological changes in the plants in relation to photosynthesis and nutrient content and that influenced the extent of herbivory, longevity and fecundity of insect pests. Elevated CO_2 resulted in slower growth rate with increased larval mortality and lowered fecundity in some lepidopteron insects such as *H. armigera* (Wu *et al.*, 2006). The life span of Japanese beetle, *Popillia japonica*, a major pest of soybean (*Glycine max*), was prolonged by 8.25 per cent when fed on foliages developed under elevated CO_2. Besides, females fed on such foliages laid approximately twice as many eggs as females fed on foliages grown under normal ambient CO_2 conditions (O'Neill *et al.*, 2008).

Insect herbivores were directly affected by changes in the quality of plant tissue exposed to elevated CO_2. The nitrogen concentration of leaves generally decreases when plants grow in elevated CO_2 and this reduces their nutritive value. In order to compensate for poor quality of food insect larvae of many species increase leaf consumption. Gypsy moth (*Lymantria dispar*) larvae that were fed on quaking aspen (*Populus tremuloides*) leaves exposed to high CO_2 performed poorer than larvae that were fed leaves grown in ambient CO_2 (Cannon, 1998). However, the response of these larvae to red oak (*Quercus rubra*) leaves illustrated that this phenomenon was host plant dependent. The moisture and CO_2 effects on insects could be noteworthy in global climate change (Hamilton, 2005). A gradual and continual rise in atmospheric CO_2 may affect a pest species directly and indirectly by influencing its host (Mahal and Aggarwal, 2010) thereby, increasing their consumption rates.

Climate change may aggravate crop pests that may cause 30-50 per cent of the yield losses in agricultural crops. Climate change in temperate regions might provide better condition for the pest to survive by stimulating a faster population build-up and recovery during spring. Further, an extension of the growing season would allow to produces larger number of generations in multivoltine species and increase the possibility of pest species invasion. Besides, infestations of potato leaf miner, *Liriomyza huidobrensis* were found to decrease, but the infestation and abundance of all other pests increased in all crops under increased temperature (Kroschel *et al.,* 2010). Changes in insect abundance and geographical extension of distribution would influence the amount of crop losses and would have an implication for plant production and food security. Major sucking pests such as aphids, whiteflies and chewing pests like cereal stem borers and common pod borers might shift to Northward direction, leading to severe damage in cereals, grain legumes, vegetables and fruit crops (Sharma, 2010).

Soybean grown under elevated CO_2 endured higher damage due to insects such as Japanese beetle, potato leafhopper, western corn rootworm and Mexican bean beetle than at ambient CO_2 (Mahal and Aggarwal, 2010) and had 57 per cent more damage. It was due to increased quantity of simple sugar levels in soybean leaves that forced the additional feeding by insects (Hamilton *et al.*, 2005). Leaves having lower nitrogen level were eaten more by insects to obtain sufficient nitrogen for their metabolism and sustained growth (Coviella and Trumble, 1999; Hunter, 2001). Elevated CO_2 severely affected the growth, fecundity and population dynamics of herbivores (Schadler *et al.*, 2007). The consumption, growth and development of phytophagous invertebrates depend strongly on the nutritional status of the plants (Mattson, 1990). The elevated CO_2 affected the nitrogen and carbohydrates in plants, which indirectly inluenced herbivores in many nutritional studies (Bezemer *et al.,* 1998b). To compensate for the reduced nitrogen and increased tannin and starch levels in the plants exposed to elevated CO_2, the consumption rate of the larvae was found to increase. Besides, there was decreased growth rate and prolongation of developmental time in leaf chewers (Lepidoptera) but increased abundance and fecundity in homoptera (Rao *et al.*, 2006).

Plants exposed to higher CO_2 levels than those to ambient CO_2 had significantly higher C: N ratios, tannin and gossypol contents and significantly low nitrogen content (Gao *et al.*, 2008). Significantly higher survival of cotton aphid occurred with increased CO_2 level, where as no significant differences were found in survival and lifetime fecundity of their predator *Propylaea japonica*. However, predator beetles experienced significantly longer stage-specific larval durations when fed with aphids from plants grown under elevated CO_2. It was speculated that under elevated CO_2, *Aphis gossypii* might become more serious pest, because of increased survival and developmental time of the aphid and the lady beetle. Castor semilooper larvae exhibited greater consumption when fed on foliage grown under elevated CO_2 than on that grown under ambient concentration. The elevated CO_2 exposed leaves were greatly digestible with higher approximate digestibility. Larval relative consumption rate increased, whereas the efficiency parameters *viz.*, relative growth rate (RGR), efficiency of conversion of ingested food (ECI) and digested food (ECD) decreased in case of larvae fed upon enriched CO_2 foliage. The consumption and weight gain of the larvae were negatively and significantly influenced by leaf nitrogen, which was found to be the most important factor affecting consumption and growth of larvae (Srinivasa Rao *et al.*, 2009).

Elevated CO_2 concentration increased the C: N ratio of plants and a high C: N ratio negatively influenced leaf chewers but did not influence the phloem- feeders. Many species of herbivorous insects would encounter lengthened larval developmental periods and greater mortality due to consumption of less nutritious host plants. The changes in population dynamics of affected species would thus influence their interactions with other insect species and plants (Coviella and Trumble1999). Host plants grown under elevated CO_2 generally had greater leaf area, biomass and C: N ratios compare to ambient CO_2 grown plants and aphids infested plants had lower biomass and leaf area than those without aphids. The various aphid species responded differently to elevated CO_2 with one species (*Myzus persicae*) increasing, one decreasing (*Acyrthosiphum pisum*), and other three being unaffected, clearly indicating that the responses were species specific. The C: N ratio was also influenced by the N content of the soil and the ability of plants and organisms to fix atmospheric N (Hughes and Bazzaz 2001). Leaf-chewing insects consumed more foliage from plants grown at elevated CO_2 concentrations than those grown at ambient CO_2. At elevated CO_2 clover shoot weight, laminar weight and laminar area were greater than at ambient CO_2. Elevated CO_2 condition did not significantly affect the population size of thrips, but laminar area scraped by pest feeding was significantly greater at elevated than at ambient CO_2 (Heagle 2003).

Elevated CO_2 levels had narcoleptic and behavioural changes in insects and projected concentrations of atmospheric CO_2 up to 1000 ppm were unlikely to affect insects directly. Secondary CO_2 induced effects through crop hosts such as digestibility, chemical defenses and canopy microclimate might affect insect fecundity or insect outbreaks (Ziska and Runion 2007).

Aphid populations could increase with increased CO_2 because of increased fecundity (Awmack *et al.*, 1996) and longer settling times (Smith, 1996). The potato aphid, *Aulacorthum solani*, responded quite differently under elevated CO_2 on bean (*Vicia faba*) and on tansy (Awmack *et al.,* 1997). Chen *et al.* (2004) correlated a higher fecundity in aphids with higher carbohydrate levels in food plants grown under elevated CO_2 level.

High concentrations of CO_2 significantly lowered the adult weight of aphid, *Myzus persicae* compared to aphids under ambient CO_2 (Himanen *et al.*, 2008). A decline in leaf miner populations under elevated CO_2 compared to ambient CO_2 treatments was also documented (Stiling and Cornelissen, 2007). Further, increase in consumption of leaves by insects was noticed when caterpillars were exposed to elevated CO_2 (Schadler *et al.,* 2007).Lincoln *et al.* (1986) and Chen *et al.* (2007b) observed that larvae of the soybean looper and cotton bollworm consumed more on their respective plants grown under elevated CO_2 than those grown under ambient CO_2, suggesting that feeding by herbivores may increase with increasing atmospheric CO_2 level. In addition, cotton bollworm also shows increased consumption on maize plants under elevated CO_2 (Yin *et al.*, 2010).

4.2 Impact of Climate Change on Natural Enemies

Climate change affects the distribution, phenology, susceptibility and inter-relationship of insects drastically, which emphasizes the risk for sustainable crop protection by losing the control on pests by their natural enemies (Sun *et al.*, 2011). Increased atmospheric CO_2 levels have an indirect influence on natural enemy fitness through changes brought in plants and their herbivore hosts. The plants grown under elevated CO_2 had high C: N ratios in tissues, which results in slower growth and developmental rates of insects and thereby increasing in the length of life stages and rendering them vulnerable to attack of natural enemies (Coviella and Trumble, 1999). Abrupt environmental changes as induced by current climatic variability are likely to exert greater influence on pests and natural enemies than the gradual climate change. Changing climatic condition may lead to ecological imbalances and disruptions due to different reaction of pests and natural enemies in their trophic system (Hance *et al.*, 2007). Atmospheric CO_2 levels may affect the performance of natural enemies and/or susceptibility of prey *via* variety of indirect effects, which potentially make prey more susceptible to their enemies. Herbivores may be physically weakened while feeding on poor hosts under elevated CO_2 conditions, and will thus be less able to defend themselves against predators and parasitoids. Enriched CO_2 may alter enemy-avoidance behavior as some aphids showed reduced responses to alarm pheromones under elevated CO_2, potentially making them more susceptible to enemy attack (Awmack *et al.*, 1997). Such effects would increase the susceptibility of herbivores to natural enemies, reducing herbivore population size under elevated CO_2 conditions (Coll and Hughes 2008).

Climate change was likely to alter the balance between insect pests, their natural enemies and their hosts and alter the balance between host and insect pest development. The predicted rise in temperature would generally favour insect development and winter survival, and rising atmospheric CO_2 concentrations may lead to a decline in food quality for plant feeding insects because of reduced foliar nitrogen levels (Evans *et al.*, 2002). It could dissociate predator-prey relationships because of a higher sensitivity of higher trophic levels to climatic variability of different temperature optima compared with pests (Voigt *et al.*, 2003). Chen *et al.* (2004) reported that elevated CO_2 increased the density and parasitism rates of wheat aphid (*Sitobium avenae*) parasitoid (*Aphidius avenae*) and reduced the mean relative growth rates and aphid consumption of its predator *Harmonia axyridis*. Vuorinen *et al.* (2004) observed that predator bug, *Podisus maculiventris* was unable to distinguish the infested plants under elevated CO_2 which might influence the searching behaviour of this predator.

In China, researchers found that when spiders moved into an area of planthopper infestation, they reduced the pest predator ratio from 9:1 to 1.5:1 within 10 days in one study area and from 5:1 to 0.03:1 within 5 days in another. Implementation of conservation practices increased spider density, limited frequency of pesticide application, the need for chemical plant protection decreased as much as 80 per cent with no measurable loss in rice yield (Riechert, 1991). Kiritani (2006) believed that climate change might decrease the abundance of *Plutella xylostella* and increase of *Trichoplusia ni* and *H. armigera*. In general, global warming may support natural enemies by increasing more number of generations than their host species. Therefore, it is expected that utilizing natural enemies would become a major control tactic in the future. Tri-trophic interaction experiment under 750 ppm CO_2concentration involving wheat, *H. armigera* larvae and its larval parasitoid *Microplitis mediator* exhibited that there was no significant change in parasitisation rate by *M. mediator* (Yin *et al.*, 2009). Similarly, Schafellner and Schopf (2008) observed that gypsy moth, *Lymantria dispar* grown under different concentration of elevated CO_2 did not affect the development of its parasitoid wasp, *Glyptapanteles liparidis* compared to ambient CO_2. Chen *et al.* (2005) observed that cotton aphid, *Aphis gossypii* experienced significant higher fecundity under elevated CO_2 for successive generations. This suggested that elevated CO_2 altered the aphid quality through changes in plant quality that certainly influenced the coccinellid predator, *Leis axyridis via* the food chain. Under elevated CO_2, coccinellid larvae had significantly higher mean relative growth rates. In addition, lady beetle fed on aphids from elevated CO_2 consumed more aphids with significantly shorter larval and pupal durations. It was concluded that under higher CO_2 condition lady beetle could alter their prey preference that might improve the biological control of aphids by coccinellids.

Stireman *et al*. (2005) compared the interaction of caterpillar and their parasitoid across a broad gradient of climatic variability and found decreased levels of parasitism as climatic variability increased. It was found that ability of parasitoids to track host populations was impaired by climatic variability. As parasitoids play an important role in regulating host populations in cropping systems, they predicted an abrupt increase in the intensity and frequency of insect outbreaks through an altered parasitoid–host dynamics as climates became more variable. Effects of elevated CO_2 on bollworm, *H. armigera* and an omnivorous bug, *Oechalia schellenbergii* revealed that bollworm larvae feeding on *Pisum sativum* grown under elevated CO_2 (700 ppm) were significantly smaller than those reared on plants grown under ambient CO_2 conditions. The bugs performed best when fed with larvae exposed to elevated CO_2 as these were smaller in size thus easier to overcome. The study concluded that generalist predators benefitted under elevated CO_2 due to increased prey vulnerability, meaning that pest species were likely to be under higher risk of predation (Coll and Hughes 2008). Interactions among the pest, light brown apple moth and its natural enemies in the future climate found to have changes in the fitness level of natural enemies in response to changes in herbivore quality and size induced by temperature and CO_2 effects on plants. The susceptibility of herbivores to predation and parasitism decreased through the production of additional plant foliage or altered timing of herbivore life cycles in response to plant phenological changes (Thompson *et al*., 2010).

Studies have shown that increased temperatures and other human-altered factors have the potential to disrupt trophic interactions between insects and plants due to species-specific responses to climate change. Effects of climate change on parasitoid and host asynchrony might be direct or indirect through changes on host plants (Klapwijk *et al*., 2010). Life history traits such as resource specificity, geographical locations, trophic level and dispersal ability are potentially good predictors of the magnitude and direction of the response of insect species to climate change. The phenotypic changes in insects have also been linked to recent climate change that are occurring both spatially and temporally, directly from changing climate and indirectly through interactions with species in lower (plant) and upper (predators) trophic levels (Pelini *et al.*, 2009). Whitefly (*Bemisia tabaci*), parasitoid (*Encarsia formosa*) parasitization rate, developmental time and adult emergence rate was not significantly affected under elevated CO_2 for three successive generations. This outcome suggested that biological control of whitefly by their parasitoid was not altered under elevated CO_2(Wang, 2009).The effect of elevated CO_2 on natural enemies is indirect and mediated through changes in herbivores that feed on plants with altered nutritional quality. The effects resulted in changes in natural enemy development, fitness, abundance and mortality (Thomson *et al.*, 2010).

5. Conclusion

Increasing CO_2 and temperature affect the herbivore quality that ultimately alters the natural enemy fitness. The susceptibility of the herbivores to the natural enemies could be decreased through dense plant foliage. Changing climate may alter the abundance and activity of natural enemies directly *via* changes in quality of herbivore and indirectly through adoption of new management strategies by farmers to cope up with climate change. These strategies may create a spatio-temporal asynchrony between pests and their natural enemies thus adversely affecting bio control efforts. Because of the multifarious effects of climate change on natural enemies, predictions about climate impacts are not so easy unless there is a good understanding of the environmental impacts on tri-trophic interactions. Climate change could affect efficacy of crop protection chemicals through changes in temperature, rainfall pattern, morphological and physiological changes in crop plants. As an adaptation to climate change, changes in package of practices of rice such as sowing time and pesticide application rates may have to be revised to effectively suppress the pest population and sustained agricultural production. Thus calls for meticulous research planning and development of suitable pest management strategies are warranted to ensure food security in future.

6. References

1. Ahmed FE, Hall AE, Madore MA. Interactive effects of high temperature and elevated carbon dioxide concentration on cowpea (Vigna unguiculata (L.) Walp. Journalof Plant Cell and Environment. 1993; 16:835-842.

2. Andrew NR, Hughes L. Diversity and assemblage structure of phytophagous Hemiptera along a latitudinal gradient: predicting the potential impacts of climate change. Global Ecology and Biogeography. 2005; 14:249-262.

3. Awmack CS, Harrington R, Leather SR. Host plant effects on the performance of the aphid *Aulacorthum solani* (Kalt.) (Homoptera: Aphididae) at ambient and elevated CO_2. Global Change Biology. 1997; 3:545-549.

4. Awmack CS, Harrington R, Leather SR, Lawton JH. The impact of elevated CO_2 on aphid-plant interactions. Aspects of Applied Biology. 1996; 45:317-322.

5. Awmack CS, Harrington WR, Lindroth RL. Aphid individual performance may not predict population responses to elevated CO_2 or O_3. Global Change Biology. 2004; 10:1414-1423.

6. Baker RHA, Sansford CE, Jarvis CH, Cannon RJC, MacLeod A, Walters KFA. The role of climatic mapping in predicting the potential geographical distribution of non-indigenous pests under current and future climates. Agriculture, Ecosystem and Environment. 2000; 82:57-71.

7. Bale JSB, Masters GJ, Hodkinson ID, Awmack C, Bezemer TM. Herbivory in global climate change research: direct effects of rising temperature on insect herbivores. Global Change Biology. 2002; 8:1-16.

8. Bezemer FA, Chiariello NR, Coley PD, Pitelka LF. Plant-insect herbivore interactions in elevated atmospheric CO_2: quantitative analyses and guild effects. Odis. 1998a; 82:212-222.

9. Bezemer TM, Jones TH, Knight KJ. Long-term effects of elevated CO_2 and temperature on populations of the peach potato aphid Myzus persicae and its parasitoid Aphidius matricariae. Oecologia. 1998b; 116:128-135.

10. Butler GD, Henneberry TJ, Clayton TE. *Bemisia tabaci* (Homoptera, Aleyrodidae) development, oviposition and longevity in relation to temperature. Annalsof Entomological Society of America. 1983; 76:310-313.

11. Cannon RJC. The implications of predicted climate change for insect pests in the UK, with emphasis on non-indigenous species. Global Change Biology. 1998; 4:785-796.

12. Charles JG, Kean JM, Chhagan A. Developmental parameters and voltinism of the painted apple moth, Teia anartoides Walker (Lepidoptera: Lymantriidae) in New Zealand. New Zealand Entomologist. 2006; 29:27-36.

13. Chen FJ, Wu G, Ge F. Impacts of elevated CO_2 on the population abundance and reproductive activity of aphid *Sitobion avenae* Fabricius feeding on spring wheat. Journal of Applied Entomology. 2004; 128:723-730.

14. Chen FJ, Wu G, Parajulee MN, Ge F. Long term impacts of elevated carbon dioxide and transgenic *Bt* cotton on performance and feeding of three generations of cotton bollworm. Entomologia Experimentalis Et Applicata. 2007b; 124:27-35.

15. Chen F, Ge F, Parajulee MN. Impact of elevated CO_2 on tri-trophic interaction of Gossypium hirsutum, Aphis gossypii, and Leis axyridis. Environmental Entomology. 2005; 34:37-46.

16. Coll M, Hughes L. Effects of elevated CO_2 on an insect omnivore: A test for nutritional effects mediated by host plants and prey. Agriculture Ecosystems and Environment. 2008; 123:271-279.

17. Coviella CE, Trumble JT. Effects of elevated atmospheric carbon dioxide on insect-plant interactions. Conservation Biology. 1999; 13(4):700-712.

18. Das DK, Singh J, Vennila S. Emerging crop pest scenario under the impact of climate change-A brief review. Journal of Agricultural Physics. 2011; 11:13-20.

19. Diku A, Mucaj L. Identification and implementation of adaptation response measures to Drini-Mati River Deltas. Report on expected climate change impacts on agriculture & livestock and their influence in the other economic sectors in the DMRD. UNDP Climate Change Program, 2010, 34.

20. Dugal A, Yelle S, Gosselin A. Influence of CO_2 enrichment and its method of distribution on the evolution of gas exchanges in green house tomatoes. Canadian Journal of Plant Science. 1990; 70:345-356.

21. El Kohen A, Pontailler JY, Mousseau M. Effect of doubling of atmospheric CO_2concentration of dark respiration in aerial parts of young chestnut trees (*Castanea sativa* Mill.) *Comptes Rendus De L* Academic Des Sciences Serie Iii- Sciences De La Vie. 1991; 312:477-481.

22. Evans H, Straw N, Watt A. Climate change: Implicat-ions for insect pests. Forestry Commission Bulletin. 2002; 125:99-118.

23. Fajer ED. The effects of enriched CO_2 atmospheres on plant-insect herbivore interactions: growth responses of larvae of the specialist butterfly, *Junonia coenia* (Lepidoptera: Nymphalidae). Oecologia. 1989; 81:514-520.

24. Flynn DFB, Sudderth EA, Bazzaz FA. Effects of aphid herbivory on biomass and leaf-level physiology of *Solanum dulcamara* under elevated temperature and CO_2. Environmental and Experimental Botany. 2006; 56:10-18.

25. Gao F, Zhu SR, Sun YC, Du L, Parajulee M, Kang L *et al*. Interactive effects of elevated CO_2 and cotton cultivar on tri-trophic interaction of Gossypium hirsutum, Aphis gossypii and Propylaea japonica. Environmental Entomology. 2008; 37(1):29-37.

26. Grononger JW, Seiler JR, Zedakar SM, Berrang PC. Photosynthetic responseof loblolly pine and sweet gum seedling stands to elevated carbon dioxide, water stress, and nitrogen level. Canadian Journal of Forest Research. 1996; 26:95-102.

27. Gutierrez AP, Ponti L, d'Oultremont T, Ellis CK. Climate change effects on poikilotherm tri-trophic interactions. Climatic Change. 2008; 87:167-192. DOI: 10.1007/s10584-007-9379-4.

28. Hall A, Allen Jr LH. Designing cultivars for climatic conditions of the next century. In: (Eds. Buxton DR. and Allen LH.). International crop science IInternational Crop science Congress. Crop Science Society of America, Madison, Wisconsin, 1993, 291-297.

29. Hamilton JG, Dermody O, Aldea M, Zangerl AR, Rogers A, Berenbaum MR *et al*. Anthropogenic changes in tropospheric composition increase susceptibility of soybean to insect herbivory. Environmental Entomology. 2005; 34:2479-2485.

30. Hance T, Baaren JV, Vernon P, Boivin G. Impact of extreme temperatures on parasitoids in a climate change perspective. Annual Review of Entomology. 2007; 52:107-126.

31. Harrington R, Fleming R, Woiwood IP. Climate change impact on insect management and conservation in temperate regions: can they be predicted? Agricultural and Forest Entomology. 2001; 3:233-240.

32. Hasegawa T, Shimono H, Yang LX, Kim HY, Kobayashi T, Sakai H *et*

al. Response of rice to increasing CO_2 and temperature: recent findings from large-scale free-air CO_2 enrichment (FACE) experiments. In: Proceedings of the 26th international rice conference 9-12 October, 2006. (Eds Aggarwal P, Ladha J, Singh R, Devakumar C, Hardy B.) New Delhi, India, 2007, 439-447.

33. Heagle AS. Influence of elevated carbon dioxide on interactions between Frankliniella occidentalis and Trifolium repens. Environmental Entomology. 2003; 32(3):421- 424.
34. Himanen SJ, Nissinen A, Dong WX, Nerg AM, Stewart CN, Poppy GM *et al*. Interactions of elevated carbon dioxide and temperature with aphid feeding on transgenic oilseed rape: Are Bacillus thuringiensis (Bt) plants more susceptible to non-target herbivores in future climate? Global Change Biology. 2003, 2008; 14:1437-1454.
35. Hughes L, Bazzaz FA. Effects of elevated CO_2 on five plant-aphid interactions. Entomologia Experimentalis Et Applicata. 2001; 99(1):87-96.
36. Hulle M, Bonhomme J, Maurice D, Simon JC. Is the life cycle of high arcticaphids adapted to climate change? Polar Biology. 2008; 31:1037-1042.
37. Hunter MD. Effects of elevated atmospheric carbon dioxide on insect-plant interactions. Agriculture and Forest Entomology. 2001; 3:153-159.
38. IPCC. Summary forpolicymakers, 2001, 1.7, http://www.ipcc.ch/.IPCC.
39. IPCC. Summary for policymakers. In: Climate change the physical science basis. Contribution of Working Group I to the Fourth Assessment Report of the Intergovernmental Panel on Climate Change. Solomon S, Qin D, Manning M, Chen Z, Marquis M, Averyt KB *et al*. (Eds). Cambridge: Cambridge University Press, 2007, 1-18.
40. Johnson SN, Lopaticki G, Hartley SE. Elevated atmospheric CO_2 triggers compensatory feeding by root herbivores on a C_3 but not a C_4 grass. PLoS ONE. 2014; 9(3):1-5. e90251. doi:10.1371/journal.pone.0090251.
41. Kiritani K. Predicting impacts of global warming on population dynamics and distribution of arthropods in Japan. Population Ecology. 2006; 48:5-12.
42. Klapwijk MJ, Grobler C, Ward K, Wheeler D, Lewis O. Influence of experimental warming and shading on host-parasitoid synchrony. Global climate Change Biology. 2006; 16:102-112.

43. Kroschel J, Sporleder M, Simon R, Juarez H, Gonzales J, Carhuapoma P *et al*. Predicting the effects of global warming on insect pests. Technical innovation brief, 2010, 5. Www.ucipm.edu. accessed on 30/2/2013.

44. La Marche VC, Grabyll DA, Fritis HC, Rose MR. Increasing atmospheric carbon dioxide: tree ring evidence for growth enhancement in natural vegetation. Science. 1984; 225:1019-1021.

45. Lawler IR, Foley WJ, Woodrow IE, Cork SJ. The effects of elevated CO_2atmospheres on the nutritional quality of eucalyptus foliage and its interactions with soil nutrient and light availability. Oecologia. 1997; 109:59-68.

46. Lewis T. Thrips as crop pests. CAB International, Cambridge: University Press, 1997, 740.

47. Lincoln DE, Couvet D, Sionit N. Response of an insect herbivore to host plants grown in carbon dioxide enriched atmospheres. Oecologia, 1986; 69:556-560.

48. Long SP, Ainsworth EA, Leakey ADB, Nosberger J, Ort DR. Food for thought: lower-than-expected crop yield stimulation with rising CO_2 concentrations. Science. 2006; 312:1918-1921.

49. Mahal MS, Agarwal N. Impact of Global climate change on arthropod fauna, In: Souvenier: National symposium on perspectives and challenges of integratedpest management for sustainable agriculture. Dr. YS Parmer university of Agriculture and forestry, Nauni, Solan, India, November, 2010, 19-21, 50-56.

50. Marks S, Lincoln DE. Antiherbivore defense mutualism under elevated carbon dioxide levels: a fungal end ophyte and grass. Environmental Entomology. 1996; 25:618-623.

51. Mattson WJ. Jr. Herbivory in relation to plant nitrogen content. Annual Reviews of Ecology and Systematics. 1990; 11:119-161.

52. Memmott J, Craze PG, Waser NM, Price MV. Global warming and the disruption of plant–pollinator interactions. Ecology Letters. 2007; 10:710-717. (doi:10. 1111/j.1461-0248.2007.01061.x)

53. Mitchell RAC, Mitchell VJ, Dirscoll SP, Franklin I, Lawlor DW. Effects of increased carbon dioxide concentration and temperature on growth and yield of winter wheat at two levels of nitrogen application. Plant Cell and Environment. 1993; 16:521-529.

54. Miyagi KM, Kinugasa T, Hikosaka K, Hirose T. Elevated CO_2

concentration, nitrogen use, and seed production in annual plants. Global Change Biology. 2007; 13:2161-2170.

55. Netherer S, Schopf A. Potential effects of climate change on insect herbivores in European forests- General aspects and the pine Processionary moth as specific example. Forest Ecology and Management. 2010; 259(4-5):831-838.

56. Neumeister L. Climate change and crop protection: anything can happen. Published by PAN Asia and the Pacific, 2010, 41.

57. O'Neill BF, Zangerl AR, De Lucia EH, Berenbaum MR. Longevity and fecundity of Japanese beetle (*Popillia japonica*) on foliage grown under elevated carbon dioxide. Environmental Entomology. 2008; 37(2):601-607.

58. Osbrink WLA, Trumble JA, Wagner RE. Host suitability of *Phaseolus lunatus*for *Trichoplusiani* (Lepidoptera: Noctuidae) in controlled carbon dioxide atmospheres. Environmental Entomology. 1987; 16:639-644.

59. Parmesan C. Influences of species, latitudes and methodologies on estimates of phenological response to global warming. Global Change Biology. 2007; 13:1860-1872. (doi:10.1111/j.1365-2486.2007.01404.x).

60. Pelini SL, Prior KM, Parker DJ, Dzurisin JDK, Lindroth RL, Hellmann JJ. Climate change and temporal and spatial mismatches in insect communities, pp. In: (Ed. Letcher, T), Climate Change: Observed Impacts on Planet Earth. Elsevier BV. limited, Netherlands, 2009, 215-230.

61. Peng S, Huang J, Sheehy JE, Laza RC, Visperas RM, Zhong X *et al.* Rice yields decline with higher night temperature from global warming. Proceedings of National Academy of Sciences. 2004; 101(27):9971-9975.

62. Piikki K, Vorne V, Ojanpera K, Pleijel H. Impact of elevated O_3 and CO_2 exposure on potato (*Solanum tuberosum* L. cv. Bintje) tuber macronutrients (N, P, K, Mg and Ca). Agriculture Ecosystems and Environment. 2007; 118:55-64.

63. Poorter H, Pot S, Lambers H. The effect of an elevated atmospheric carbon dioxide concentration on growth, photosynthesis and respiration of Plantago major. Physiologia Plantarum. 1988; 73:553-559.

64. Rao MS, Khan MAM, Srinivas K, Vanaja M, Rao GSN, Ramakrishna YS. Effects of Elevated carbon dioxide and temperature on insect- plant

interactions-A review. Agricultural Review. 2006; 27(3):200-207.

65. Reddy AR, Rasineni GK, Raghavendra AS. The impact of global elevated CO_2 concentration on photosynthesis and plant productivity. Current Science. 2010; 99(10):46-57.

66. Reekie EG, Bazzaz FA. Phenology and growth in four annual species grown in ambient and elevated CO_2. Canadian Journal of Botany. 1991; 69:2475-2481.

67. Riechert SE. Prey abundance versus diet breadth in a spider test system. Evolutionary Ecology. 1991; 5:327-338.

68. Root TL, Schneider SH. Can large-scale climatic models be linked with multiscale ecological studies? Conservation Biology. 1993; 7:256-270.

69. Schadler M, Roeder M, Brandal R. Interacting effects of elevated CO_2, nutrient availability and plant species on a generalist invertebrate herbivore. Global Change Biology. 2007; 13:1005-1015.

70. Schafellner C, Schopf A. The greenhouse diet: gypsy moth performance in a CO_2-enriched world. Mitteilungen der Deutschen Gesellschaft fu¨r allgemeine und angewandte Entomologie. 2008; 16:71-75.

71. Sharma HC. Global warming and climate change: Impact on arthropod biodiversity, pestmanagement, and food security. In: Souvenier: National symposium on perspectives and challenges of integrated pest management for sustainable agriculture. Dr. YS Parmer university of Agriculture and forestry, Nauni, Solan, India. 2010; 19-21, 3-14.

72. Smith H. The effects of elevated CO_2 on aphids. Antenna. 1996; 20:109-111. Chen FJ, Wu G, Ge F. Impacts of elevated CO_2 on the population abundance and reproductive activity of aphid *Sitobion avenae* Fabricius feeding on spring wheat. Journal of Applied Entomology. 2004; 128(9-10):723-730.

73. Srinivasa Rao M, Srinivas K, Vanaja M, Rao GSN, Venkateswarlu B, Ramakrishna YS. Host plant (*Ricinus communis* Linn.) mediated effects of elevated CO_2 on growth performance of two insect folivores. Current Science. 1996; 97:1047-1054.

74. Stiling P, Cornelissen T. How does elevated carbon dioxide (CO_2) affect plant-herbivore interactions? A field experiment and meta-analysis of CO_2-mediated changes on plant chemistry and herbivore performance. Global Change Biology. 2007; 13:1823-1842.

75. Stireman JO, Dyer LA, Janzen DH, Singer S, Lill JT, Marquis JR. et al.

Climatic unpredictability and parasitism of caterpillars: implications of global warming. Proceedings of the National Academy of Sciences. 2005; 102:17384-17386.

76. Sun YC, Feng L, Gao F, Ge F. Effects of elevated CO_2 and plant genotype on interactions among cotton, aphids and parasitoids. Insect Science. 2011; 18:451-461.

77. Thomson LJ, Macfadyen S, Hoffmann AA. Predicting the effects of climate change on natural enemies of agricultural pests. Biology. 2010; 52:296-306.

78. Trumble JT, Kolodny HDM, Ting IP. Plant compensation for arthropodherbivory. Annual Review Entomology. 1993; 38:93-119.

79. Voigt W, Perner J, Davis AJ, Eggers T, Schumacher J, Bahram R. Impact of climate change on crop-pest and pest-natural enemy interactions. Ecology. 2003; 84:2444-2453.

80. Vuorinen T, Nerg AM, Ibrahim AM, Reddy GV, Holopainen JK. Emission of *Plutella xylostella* induced compounds from cabbages grown at elevated CO_2 and orientation behavior of the natural enemies. Plant Physiology. 2004; 135:1984-1992.

81. Wang KH, Tsai JH. Temperature effect on development and reproduction of silver leaf whitefly (Homoptera: Aleyrodidae). Annals of Entomological Society ofAmerica. 1996; 89:375-384.

82. Wang XX. Impacts of elevated CO_2 on systems of cotton – *Bemisia tabaci* biotype B–*Encarsia formosa*. Master thesis, Fujian Normal University, China, 2009.

83. Watt AD, Whittaker JB, Docherty M, Brooks GL, Lindsay E, Salt DT. The impact of elevated CO_2 on insect herbivores. In: Harrington, R. and Stork, N. E: Insects in a Changing Environment, 17th Symposium of the Royal Entomological Society. Academic Press, London, 1995, 197-217.

84. Whittaker JB. Impacts and responses at population level of herbivorous insects to elevated CO_2. Environmental Entomology. 1999; 96:149-156.

85. Wu G, Chen FJ, Ge F. Direct effects of elevated CO_2 on growth, development and reproduction of cotton bollworm Helicoverpa armigera Hubner. Acta Ecologica Sinica. 2006; 25(6):1732-1738.

86. Yamamura K, Kiritani K. A simple method to estimate the potential increase in the number of generations under global warming in temperate zones. Applied Entomology and Zoology. 1998; 33:289-298.

87. Yin J, Sun Y, Wu G, Ge F. Effects of elevated CO_2 associated with maize, a C_4 plant, on multiple generations of cotton bollworm Helicoverpa armigera Hubner (Lepidoptera: Noctuidae). Entomologia Experimentalis Et Applicata. 2010; 136:12-20.

88. Yin J, Sun Y, Wu G, Parajulee MN, Ge F. No effects of elevated CO_2 on the population relationship between cotton bollworm, *Helicoverpa armigera* Hübner (Lepidoptera: Noctuidae), and its parasitoid, Microplitis mediator Haliday (Hymenoptera: Braconidae). Agriculture Ecosystem and Environment. 2009; 132:267-275.

89. Zhang G, Sakai H, Tokida T, Usui Y, Zhu J, Nakamura H *et al*. The effects of free-air CO_2 enrichment (FACE) on carbon and nitrogen accumulation in grains of rice (*Oryzasativa* L.). Journal of Experimental Botany. 2013; 64(11):3179-3188.

90. Ziska LH, Teramura AH. CO_2 enhancement of growth and photosynthesis in rice (Oryza sativa): Modification by increased Ultraviolet-B radiation. Plant Physiology. 1992; 99:473-481.

91. Ziska LS, Runion GR. Future weed, pest and disease problems for plants. In: Agroecosytems in a changing climate. (Eds. Newton CD, Carran RA, Edwards GR and Niklaus PA). Taylor and Francis. UK, 2007, 261-273.

Chapter - 2
Recent Trends in Detection and Diagnosis of Plant Diseases

Authors

Raghu S.
Crop Protection Division, ICAR-National Rice Research Institute (NRRI), Cuttack-753006, Odisha, India

Kutubuddin Ali Molla
Crop Protection Division, ICAR-National Rice Research Institute (NRRI), Cuttack-753006, Odisha, India

Guru Prasanna Pandi G.
Crop Protection Division, ICAR-National Rice Research Institute (NRRI), Cuttack-753006, Odisha, India

Prabhukarthikeyan S.R.
Crop Protection Division, ICAR-National Rice Research Institute (NRRI), Cuttack-753006, Odisha, India

Mathew S. Baite.
Crop Protection Division, ICAR-National Rice Research Institute (NRRI), Cuttack-753006, Odisha, India

Manoj K Yadav
Crop Protection Division, ICAR-National Rice Research Institute (NRRI), Cuttack-753006, Odisha, India

Mayabini Jena
Crop Protection Division, ICAR-National Rice Research Institute (NRRI), Cuttack-753006, Odisha, India

Chapter - 2

Recent Trends in Detection and Diagnosis of Plant Diseases

Raghu S, Kutubuddin Ali Molla, Guru Prasanna Pandi G, Prabhukarthikeyan S.R, Mathew S Baite, Manoj K Yadav and Mayabini Jena

1. Introduction

Plant diseases are the most important biotic factors which have a direct impact on global agricultural productivity and food security under changing climatic scenario. It is estimated that the combined infestation of pest and diseases in crop plants could take a heavy toll up to 50% losses in attainable yield in many crops especially major food crops like rice, wheat and maize. The losses may be at field condition (Pre-harvest losses) and at storage (Post-harvest losses). If we combine both pre and post harvest losses, the result would much more as storage losses leads to deterioration of quality which is a major concern in many countries of the world (Oerke, 2006; Gautam *et al.* 2013). Plant diseases alone can cause almost 20% reductions in yield in major food and cash crops worldwide. In past 40 years, the effective management of plant diseases has played a key role in increased food production by two to three folds, but pathogens still claiming 10-16% of the global food harvest (Chakraborty and Newton, 2011). In addition global climate change also posing a serious threat to crop production where many major diseases become still aggressive and minor diseases becoming major problem. A successful plant disease management mainly depends on accurate, efficient and quick detection of plant pathogens from a symptomatic plant. Thereafter, studying the exact causes of the illness, pathogen responsible for it, amount of genetic and pathogenic variability in the pathogen population followed by development of disease resistant varieties along with integrated management practices.

2. Modern Agriculture and Plant Pathology

The demand of food production to ever increasing global population is increasing at rapid rate rising concern about global food security. The progress in food security is achieved through modern agriculture methods like intensive monoculture of high yielding varieties and hybrids which demand large amount of inputs such as fertilizer and water to maximize productivity per unit area. This rapid change due to green revolution changed from small-scale, single-cycle annual and diversified production systems in traditional agriculture to high-intensity, multi-cropping, monoculture production systems. This has disrupted the co evolutionary dynamics that are representatives of plant-pathogen interactions in natural ecosystems (Mundt *et al.* 2011; Sommerhalder 2011). Modern agriculture created unique conditions for the rapid development of highly infective pathogen populations with increased frequency of epidemics. New pathogenic races/isolates were evolved as a result of random mutation and recombination causing much more damage compared to earlier. This also created opportunity for the pathogens that shift from a parasitic to saprophytic mode of survival in off-season and become pathogenic again in crop season (Altizer *et al.* 2006; Sommerhalder *et al.* 2011). Thus accurate and early detection of plant pathogens from an infected plant is most important step in prevention of losses. There are various detection and diagnostic methods useful in plant disease detection. Some of them are discussed in this chapter.

3. Markers in Plant Disease Detection and Diagnosis

3.1 Biochemical Markers

Biochemical markers are proteins produced by gene expression. These proteins can be isolated and identified by gel electrophoresis and staining. Isozymes are proteins that catalyze the same enzymatic reaction; they are revealed on electrophoregrams through a colored reaction associated with the enzymatic activity. They are the product of the various alleles of one or several genes. Monomeric and dimeric isozymes are most often used because the analysis of their segregation is easier. Isozymes are generally codominant. Examples of isozyme systems routinely used in plant breeding for disease resistance: isozyme *Aps-1* for its linkage to the tomato resistance to nematode (*Meloidogyne* spp.) from *Lycopersicon peruvianum* (Rick and Fobes, 1974); isozyme *Got-2* for its linkage to the tomato resistance to *Fusarium oxysporum* race 3 from *L. pimpenellii* (Bournival *et al*, 1989); isozyme *Est-5* for its linkage to the wheat resistance to *Meloigogyne naasi* from *Aegilops variabilis* (Yu, 1991). It should be noted that, in most cases, the polymorphism of isozyme markers is rather poor within a cultivated species (eg, tomato). This explains why most of the known cases originate from interspecific crosses in which the resistance gene(s) come(s) from a wild relative.

3.2 DNA Based Markers

Different criteria or parameters have been followed to classify the DNA marker systems Based on the chronology of their development, DNA markers are classified as (1) first-generation (RFLP, RAPD, and their modifications), (2) second-generation (SSRs, AFLPs, and their modifications), and (3) third-generation (ESTs and SNPs) markers (Singh and Singh, 2015). They can also be classified as a) PCR based and b) non-PCR based, depending on the employment of PCR. And also as (1) SNPs (generated by variation in DNA sequence) and (2) non-SNPs (produced by variation in sequence length, e.g. SSRs) based on their molecular basis (Gupta *et al.* 2001). Another classification approach uses the location and the functional significance of markers. On this basis, the markers are grouped as (1) random, (2) gene-based, and (3) functional markers. But from the user's point of view, a more useful classification of markers would be the one based on the method of marker detection and genotyping since this would indicate the degree of their user-friendliness. On the basis of the above and the throughput criteria, the various marker systems can be grouped as (1) low-throughput hybridization-based markers, (2) medium-throughput PCR-based markers, and (3) high-throughput sequence-based markers. This grouping is often considered as (1) first-generation, (2) second-generation, and (3) third-generation molecular markers, respectively.

3.3 PCR Based Markers

The polymerase chain reaction (PCR) uses the property of DNA to melt or denaturate when heated in a solution (i.e. the separation of 2 strands). This provides the opportunity for copying each strand through the operation of a DNA polymerase enzyme adding nucleotides to form the complementary strand starting from sites at which primers have recognized complementary sequences in the template DNA. A thermo cycler is used to vary the temperature in the reaction tubes according to optimal specified cycles. Primers can be randomly synthesized sequences (typically decamers) or sequences determined from a known sequence. Each cycle doubles the quantity of DNA copied from the small amount present at the start, thus exponentially amplifying the DNA sequence located between the 2 known DNA primers by copying it many times (20-45 cycles of amplification) leading to a sufficient quantity of copies for direct detection after electrophoresis.

Table 1: Markers in characterization of plant pathogens

Pathogen	Crop	DNA profiling system	Objective	Reference
Fusarium oxysporum f.sp. *cubense*	Banana	DAF	Genetic diversity among the isolates	Bentley and Bassam, 1996
Colletotrichum graminicola	Sorghum	RAPD	Genetic diversity among the isolates	Gutherie *et al.*, 1992
Xanthomonas oryzae pv *oryzae*	Rice	DNA fingerprinting	Population structure over locations and seasons in Philippines and China	Ardales *et al.*, 1996
Piricularia grisea	rice	DNA fingerprinting	Analysis of population structure in Philippines Lineage diversity in North-China Population structure in USA over 30 years	Chen *et al.*, 1995 Zhu *et al.*, 1995 Xia *et al.*, 1993
Phytopthora infestans	Potato	DNA fingerprinting	Genetic variation in Poland Genetic Diversity in Northern and Central Mexico	Sujkowaski *et al.*, 1994
Stagnospora nodorum	Wheat	DNA fingerprinting	Genetic structure of the field population	McDonald *et al.*, 1994
Cladosporium fulvum	Tomato	AFLP	Inter and Intraspecific variation	Majer *et al.*, 1996
Alternaria brassicae	Crucifers	RAPD	Genetic variability of isolates collected from different geographical regions	Sharma and Tewari, 1995
Xanthomonas campestris pv *vesicatoria*	Pepper and Tomato	ERIC	Differentiation of Genomic structure	Louws *et al.*, 1995
Ascochyta rabei	Chickpea blight	RAPD	To assess pathotype diversity	Udupa *et al.*, 1998
Rhadopolus similis	Banana	RAPD	Molecular diversity in Srilankan population	Hahn *et al.*, 1994
Globodera palida	Potato	RFLP	Differentiation of closely related	Schinick *et al.*, 1990

			population	
Venturia inaequalis	Apple	SSR	Genetic diversity among the isolates	Kamyak *et al.*, 2016
Fusarium oxysporum f.sp.*lycopersici*	Tomato	ISSR	Genetic diversity among the isolates	Nirmaladevi *et al.*, 2016
Fusarium solani	Chilli	ITS	Genetic diversity among the isolates	Raghu., 2014
Ascochyta rabei	Chick pea	STMS	Population structure analysis	Phan *et al.*, 2003
Xanthomons oryze pv. *oryzae*	Rice	ERIC-PCR	Genetic diversity studies	Keshavarz *et al.*, 2011
Colletotrichum spp.	Onion	ITS	Genetic diversity	Patil *et al.*, 2013
Fusarium fujikuroi	Rice	SSR	Genetic diversity of isolates	Valente et al., 2017
Fusarium oxysporum	Lamb's lettuce	IGS	Molecular characterization	Srinivasan *et al.*, 2010
Fusarium spp.		Beta tubulin, ITS, IGS, rDNA RFLP, UP-PCR, RAPD, RFLP, rDNA sequencing	Whole genome, rDNA ITS, IGS region	Yli- Mattila *et al.*, 1997a, b, 2002
F. o. f. sp. *cucumerium*		RAPD, RFLP	Whole genome	Wang *et al.*, 2001
F. culmorum		Microsatellites	Whole genome	Giraud *et al.*, 2002

3.4 Markers Application in Pathogen Population Study

Three major aspects of pathogen population structure have been studied in relation to their migration, selection, genetic drift and mode of reproduction. These are a) amount of diversity in a population, b) the genetic relationship among the isolates and c) partitioning of diversity in time and space. The most informative studies have used hierarchical sampling schemes for partitioning of the total genetic diversity into several levels. The scale of sampling has been as small as several isolations from a single lesion on a leaf as large as collections from different regions of a country or continent or even from different continents. High degree of genetic diversity has been reported on a very fine spatial scale in several cases (Mohapatra *et al.*, 2003).

Different aspects of pathogen population structure have implications in breeding for varieties with more durable resistance. For instance, a genotype may be considered resistant although it is susceptible to the existing pathogen sub-populations that are not represented in the screening nursery used for variety evaluation. Such a variety, identified as resistant to a pathogen and released for commercial cultivation, will obviously break down rapidly. Thus requiring immediate replacement. In contrast, varieties showing resistance to isolates representing all the sub-populations of a pathogen, collected from divergent host genotypes over a significant period of time and space will be more durable. Break-down of resistance in such varieties will await pathogen evolution with respect to virulence. It is, in this context, essential that the population of all the important plant pathogens is thoroughly characterized using molecular markers on broad special and temporal scales (Mohapatra *et al.*, 2001).

Every host plant expresses two types of resistance reaction against pathogens, either horizontal or vertical. The vertical resistance is also called as gene-for-gene system and shows simple Mendelian pattern of monogenic resistance. The horizontal resistance, on the other hand, is polygenically inherited. Both categories of host have been characterized by employing molecular markers in many crop species by different workers.

Table 2: Molecular mapping of genes for disease resistance in crop plants

Crop	Disease	Pathogen	Gene	Marker (s)	Reference
Rice	Blast	*Pyricularia oryzae*	Pi-ta	RFLP	Yu *et al.*, 1991
			Pi-5 (t)	RFLP	Wang *et al.*, 1994
			Pi-Z^{-6}	RFLP	Mew *et al.*, 1994
			Pi-12 (t)	RFLP	Zheng *et al.*, 1995
			Pi k^{h}	RAPD	Shankar, 2002
			Pi k^{h}	AFLP	Singh, 2002
	Bacterial leaf bilght	*Xanthomonas oryzae pv oryzae*	Xa-1, xa-3	RFLP	Yoshimura *et al.*, 1992
			Xa-5	RFLP	McCouch *et al.*, 1991
			Xa-21	RFLP	Ronald *et al.*, 1992
Wheat	Powdery mildew	*Erysiphe graminis pv tritici*	Pm 24		Helgurea *at al.*, 2000
	Leaf rust	*Puccinia recondita tritici*	*Lr1*	RFLP/STS	Qiu *et al.* (2007)
			Lr9	RAPD/STS RFLP	Gupta *et al.* (2005)
			Lr10	RFLP/STS, STS	Stepien *et al.* (2003)
			Lr12	SSR	Singh and Bowden

					(2011)
			Lr13	RFLP, SSR	Bansal *et al.* (2008)
			Lr14	SSR	Herrera-Foessel *et al.* (2007)
			Lr14a	SNP	Terracciano *et al.* (2013)
			Lr15	SSR	Dholakia *et al.* (2013)
			Lr21	RFLP; KASPar	Neelam *et al.* (2013)
	Stem rust	*Puccinia graminis tritici*	*Sr25*	STS	Liu *et al.* (2010)
			Sr26	STS	Mago *et al.* (2005)
			Sr35	SSR	Babiker *et al.* (2009)
			Sr36	SSR	Tsilo *et al.* (2008)
			Sr32	SSR	Mago *et al.* (2013)
			Sr45	SSR /AFLP	Periyannan *et al.* (2014)
	Yellow rust	*Puccinia striformis*	*Yr5*	STS	Chen *et al.* (2003)
			Yr10	STS	Singh *et al.* (2009)
Maize	Turcicum leaf blight	*Helminthosporium turcicum*	*Ht-1*	RFLP	Bentolia *et al.*, 1991
	Maize dwarf virus disease	*Maize dwarf mossaic virus*	*Mdm-1*	RFLP	McMullen *et al.*, 1989
Tomato	Leaf mould	*Cladosporium fulvum*	*cfa*	RFLP	Vanderbeek *et al.*, 1992
	Fusarium wilt	*Fusarium oxysporum*	I_2	RFLP	Sarfatti *et al.*, 1989
	Bacterial leaf spot of tomato	*Pseudomonas syringe*	*Pto*	RFLP	Debener *et al.*, 1991
	Root knot	*Meloidogyne* sp	*Mi*	RFLP	Messeguer *et al.*, 1991
Potato	Potato virus X	Potato virus X	Rx_1, Rx_2	RFLP	Ritter *et al.*, 1991
Soybean	Blight	*Phytopthora sojai*	Rps 1	RFLP	Polzin *et al.*, 1994

Plants are resistant to certain pathogens because they belong to taxonomic groups that are outside the host range of these pathogens (non-host resistance), because they possess genes for resistance (R genes) directed against the virulent genes of pathogen (true, race specific, cultivar-specific, or gene-for-gene resistance), or because, for various reasons, the plants escape or tolerate infection by these pathogens (apparent resistance) (Agriose, 2005).

4. Loop-Mediated Isothermal Amplification (LAMP) Assay in Plant Pathogen Detection

Loop mediated isothermal amplification was developed by Eiken Chemical Co. Ltd. and first time described by Notomi *et al.* (2000). Loop-mediated isothermal amplification (LAMP) is a simple method of amplification of DNA from a symptomatic plant. It is a simple, cost effective and rapid method for specific on-site detection of genomic DNA by the use of six oligonucleotide primers with eight binding sites hybridizing specifically to different regions of genes, and a thermophilic DNA polymerase from *Geobacillus stearothermophilus* for DNA amplification (Nelson and Vogel, 2010; Abd-Elsalam *et al.* 2011). In LAMP technique, diagnostics can be performed without examination of the amplification product during or after amplification. Here nucleic acid is amplified under isothermal conditions at 65 °C with high specificity and amplification efficiency. Here, the positive reactions will show bright green fluorescence, while a negative reaction would remain light orange. This is the expected result when SYBR Green I dye is added, which indicate that LAMP amplicons will be visualized directly and seen with the naked eye or through ultraviolet trans-illuminator (Khiyami *et al.* 2014). This is the most useful technique to detect plant pathogenic fungi such as *Phytopthora sojae* (Dai *et al.* 2012), *Aspergillus flavus* and *A. parasiticus* (Luo *et al.* 2012) and some toxigenic *Fusarium* spp. (Abd-Elsalam *et al.* 2011; Niessen and Vogel, 2010; Niessen *et al.* 2012; Denschlag *et al.* 2012).

Table 3: Plant pathogens detected using LAMP

S. No.	Pathogen	Host
1	Tomato yellow leaf curl virus	Tomato
2	Japanese yam mosaic virus	Yam
3	Potato virus Y	Potato
4	*Erwinia amylovora*	Apple
5	*Ralstonia solanacearum*	Many host
6	*Rodococcus fascians*	Ornamentals
7	*Botrytis cinerea*	Many host
8	Squash leaf curl virus	Squash
9	Phytoplasma	Many host
10	*Fusarium fujikuroi, F.proliferatum*	Rice
11	*Fusarium oxysporum* f. sp. *ciceris*	Chickpea

5. DNA Microarray Technology

Microarray technology evolved from southern blotting where fragmented DNA is attached to a substrate and then probed with a known gene or fragment (Southern, 1975). DNA microarray (also commonly known as gene chip, DNA chip or biochip) is a collection of microscopic DNA spots attached to a solid surface. Scientists use DNA microarrays to measure the expression levels of large numbers of genes simultaneously. Here, each DNA spot contains picomoles (10^{-12} moles) of a specific DNA sequences known as probes (reporters). There are two types of microarrays a) cDNA microarray and b) oligonucleotide array (Lezar and Barros, 2010). Array technology will be useful in genotyping, mapping, genome wide identification of transcript, isoform variation between individuals, diagnostic gene profiles (e.g., disease vs healthy plants), gene discovery and functional analysis, assessment of an organism's physiological state and identifying interacting proteins/pathways (Barbara and Valent, 2011).

6. RNA Interference in Plants

RNA silencing is a homology-based process that is triggered by double-stranded RNA (dsRNA), which leads to the suppression of gene expression. It was initially discovered in plants and was thought to function as part of a defense mechanism against viruses. Subsequently it was shown to be a ubiquitous silencing mechanism that is present in all eukaryotes including protozoa, plants and animals. The term RNA interference (RNAi) was first coined for the phenomenon when it was observed in the nematode *Caenorhabditis elegans*. RNA silencing works on at least three different levels in plants: (i) cytoplasmic silencing by dsRNA results in cleavage of mRNA and is known as post-transcriptional gene silencing (PTGS); (ii) endogenous mRNAs are silenced by micro-RNAs (miRNAs), which negatively regulate gene expression by base-pairing to specific mRNAs, resulting in either RNA cleavage or arrest of protein translation; (iii) RNA silencing is associated with sequence-specific methylation of DNA and the consequent suppression of transcription [transcriptional gene silencing (TGS)].

Table 4: Use of RNAi for virus resistance in plants

Name of virus	Family	Region targeted	Results	System used	Genome
Potato virus Y	Potyviridae	HC-Pro	Immunity	Potato	RNA
Mungbean yellow mosaic India virus	Geminiviridae	Bidirectional promoter	Recovery from infection	*Vigna mungo* (black gram)	DNA

(MYMIV)					
African cassava mosaic virus (ACMV)	Geminiviridae	Replication-associated protein gene	Reduced virus accumulation	Tobacco protoplast	DNA
Tomato yellow leaf curl Sardinia virus	Geminiviridae	Replication-associated protein gene	Poor resistance	Tomato	DNA
Pepper mild mottle virus (PMMoV)	Tobamoviridae	Arbitrary sequence	Block in viral infectivity	Tobacco	RNA
Tobacco etch virus (TEV)	Potyviridae	Arbitrary sequence	No viral-specific symptoms appeared	Tobacco	RNA
Alfalfa mosaic virus (AMV)	Bromoviridae	Arbitrary sequence	Recovery from infection	Tobacco	RNA
Beet necrotic yellow vein virus (BNYVV)	Benyviridae	Coat protein	Tolerance	Tobacco	RNA
Tobacco mosaic virus (TMV)	Tobamoviridae	Replication-associated protein	Inhibition of TMV replication	Tobacco	RNA

* Abbreviation: HC-PRO, helper-component proteinase gene, Source: *Trends in plant science*

7. Application of Tissue Culture in Plant Pathology

Tissue culture is an aseptic technique of growing cell, tissue, organ and whole organism on artificial nutrient medium under *in vitro* condition. The technique has long been in use to develop disease free plants and also to develop disease resistance in agri-horticultural crops. Micro propagation and meristem tip culture are the methods commonly used for the purpose of developing disease free plantlets. Meristem culture is widely used for developing virus free plants as the meristematic zone of a plant contain actively growing cells and are generally free of viruses. What makes the meristem virus free? One possible scientific explanation is the lack of suitable plasmodesmatal and vascular connection, which are normally required for viral transmission. The success of developing virus free plants largely depends on the size of the meristem tip explants; smallest the explants higher the rate to obtain disease free plants. This method was used in crops like potato (Al-Taleb *et al*., 2011), sweet potato (Alam *et al*., 2013), citrus (Goswami *et al*., 2013), sugarcane (Fitch *et al*., 2001) and many other researchers.

8. CRISPR/Cas9: Future of Plant Protection

Recently, an outstanding method of targeted gene editing, based on the adaptive immune system of *Streptococcus pyogenes*, has been developed

(Jinek *et al.*, 2012, Cong *et al.*, 2013). The method is popularly known as CRISPR (Clustered regularly interspaced short palindromic repeats)/Cas9 (CRISPR associated) system. It operates by creating DNA double strand breaks (DSBs) at target loci to stimulate genome editing via NHEJ (non-homologous end joining) or HDR (homology-directed repair). For genome editing purpose, a specific gRNA (guide RNA) and cas9 expression cassette are cloned in transformation vector. The gRNA is a ~20 bp long RNA sequence (complementary to target sequence of the genome to be edited) located within a longer RNA scaffold. The gRNA binds to DNA in the genome and the Cas9 nuclease creates nick (DSB) to the specific part of the genome. The DSB is then repaired by cellular repair mechanism which causes the mutation. The major advantage of this system is that CRISPR/Cas9 vector can be introduced in plant to create the targeted genome edits and subsequently, selection can be done for progeny plants carrying the desired edits but have lost the CRISPR/Cas9 cassette through segregation. The edited plants will be free of any foreign DNA and therefore, can be called as non-transgenic.

The CRISPR/Cas9 system can be used efficiently to mutate the susceptibility genes (discussed in 2.1.c) and the repressor of defense genes in the host plant to confer resistance to pathogens. Wang *et al.* (2014) utilized the CRISPR/Cas9 genome editing system to mutate three homoeoalleles of mlo gene, a defense repressor in hexaploid bread wheat, to develop resistance against powdery mildew. Similarly, improved blast disease resistance has been obtained in rice plant when a transcription factor gene OsERF922 was knocked out via CRISPR/Cas9 system (Wang *et al.*, 2016). In addition, the editing system can also be employed in developing resistance against DNA viruses. For example, the bean yellow dwarf virus (BeYDV) (gemini virus) genome has been targeted for destruction with the CRISPR–Cas system expressed in transgenic plants and a reduced virus load and disease symptoms were observed in the transgenic plants (Baltes, 2015). As virus enters into the host plant cell to propagate, the expression of CRISPR/Cas system targeted to the viral genome in plant is a good strategy to reduce the level of damage due to the particular viral infection.

9. Future Strategy

The future of agriculture will increasingly rely on the integration of biotechnology with traditional agricultural practices. Diseases can be caused by a variety of plant pathogens including fungi, bacteria, virus, nematodes and mycoplasma and there management requires the use of techniques including transgenics, molecular biology and genetics. These include; genes that express proteins or antimicrobial compounds that are directly toxic to pathogens or that reduce their growth *in situ*; gene products that directly inhibit pathogen virulence products or enhance plant structural defense, that directly or indirectly activate general plant defense responses; and resistance genes involved in the hypersensitive response and the interactions with virulence factors.

10. Conclusion

Considerable progress has been made in the use of traditional and advanced technologies for the accurate detection and diagnosis of plant pathogens. The ease of basic understanding of the interactions between pathogens and their host plants has been greatly enhanced. Use of molecular markers, Loop mediated isothermal amplification (LAMP) and DNA microarray technology, tissue culture and omics approaches have provided great applications in early and accurate detection of plant pathogens from symptomatic or even from non-symptomatic plants have been of useful in minimizing the losses caused by plant pathogens. But still there are many areas to be exploited to better understanding of host pathogen systems and possible role in disease outbreak. This will greatly helpful in efficient management of plant diseases.

References

1. Abd-Elsalam KA, Bahkali A, Moslem M, Amin O, Niessen L. An optimized protocol for DNA extraction from wheat seeds and loop-mediated isothermal amplification (LAMP) to detect Fusarium graminearum contamination of wheat grain. Int. J Mol. Sci. 2011; 12(12):3459-3472.

2. Agrios GN. Plant Pathology. 5th Ed., Elsevier Academic Press. New York, 2005, 213-223.

3. Alam I, Sharmin SA, Naher MK, Alam MJ, Anisuzzaman M, Alam MF. Elimination and detection of viruses in meristem-derived plantlets of sweet potato as a low-cost option toward commercialization. 3 Biotech. 2013; 3(2):153-164.

4. Al-Taleb MM, Hassawi DS, Abu-Romman SM. Production of virus free potato plants using meristem culture from cultivars grown under Jordanian environment. The American-Eurasian Journal of Agricultural and Environmental Sciences. 2011; 11(4):467-472.

5. Altize S. Seasonality and the dynamics of infectious diseases. Ecol. Lett. 2006; 9:467-484.

6. Ardales EY, Leung H, Veracruz CM, Mew TW, Leach JE, Nelson RJ. Hierarchial analysis of spatial variation of the rice bacterial blight pathogen across diverse agroecosystems in the Phillippines. Phytopathology. 1996; 86:241-252.
7. Babiker E, Ibrahim AMH, Yen Y, Stein J. Identification of a microsatellite marker associated with stem rust resistance gene Sr35 in wheat. Aust. J Crop Sci. 1996; 3:195-200.
8. Baltes NJ, Hummel AW, Konecna E, Cegan R, Bruns AN, Bisaro DM, Voytas DF. Conferring resistance to geminiviruses with the CRISPR–Cas prokaryotic immune system. Nature Plants. 2015; 1:15145.
9. Bansal UK, Hayden MJ, Venkata BP, Khanna R, Saini RG, Bariana HS. Genetic mapping of adult plant leaf rust resistance genes Lr48 and Lr49 in common wheat. Theor. Appl. Genet. 1996, 2008; 117:307-312. Doi: 10.1007/s00122- 008-0775-6.
10. Bentley S, Bassam BJ. A robust DNA amplification fingerprinting system applied to analysis of genetic variation within Fusarium oxysporum f. sp. cubense. J Phytopathol. 1996; 144:207-213.
11. Bentolia S, Guitton C, Bouvet N, Sailland A, Nykaza S, Freyssinet G. Identification of an RFLP marker tightly linked to the Ht1 gene in maize. Theor. Appl. Genet. 1991; 82:393-398.
12. Bournival BL, Scott JW, Vallejos CE. An isozyme marker for resistance to race 3 of Fusarium oxysporum f. sp. lycopersici in tomato. Theor. Appl. Genet. 1989; 78(4):489-494.
13. Chakraborty S, Newton AC. Climate change, plant diseases and food security: an overview. Plant Pathol. 1989; 60:2-14.
14. Chen D, Zeigler RS, Leung H, Chen RS. Population structure of pyricularia grisea at two screening sites in the Phillippines. Phytopathology. 1995; 85:1011-1020.
15. Chen X, Soria MA, Yan G, Sun J, Dubcovsky J. Development of sequence tagged site and cleaved amplified polymorphic sequence markers for wheat stripe rust resistance gene Yr5. Crop Sci. 2003; 43:2058-2064. doi: 10.2135/cropsci.2058
16. Cong L, Ran FA, Cox D, Lin S, Barretto R, Habib N *et al*. Multiplex genome engineering using CRISPR/Cas systems. Science. 2013; 339(6121):819-823.
17. Debener T, Lehnackers H, Arnold M, Dang JL. Identification and molecular mapping of a single Arabidopsis thalina locus determining

resistance to a phytopathogenic Pseudomonas syringe isolate. Plant J, 1991; 1:289-302.

18. Denschlag C, Vogel RF, Niessen L. Hyd5 gene-based detection of the major gushing-inducing Fusarium spp. In a loop-mediated isothermal amplification (LAMP) assay. Int. J Food Microbiol. 1991; 156(3):189-196.
19. Dholakia BB, Rajwade AV, Hosmani P, Khan RR, Chavan S, Reddy DMR. Molecular mapping of leaf rust resistance gene Lr15 in hexaploid wheat. Mol. Breed. 2013; 31:743-747. Doi: 10.1007/s11032-012- 9813-9.
20. Fitch MM, Lehrer AT, Komor E, Moore PH. Elimination of Sugarcane yellow leaf virus from infected sugarcane plants by meristem tip culture visualized by tissue blot immunoassay. Plant Pathology. 2001; 50(6):676-80.
21. Gautam HR, Bhardwaj ML, Kumar R. Climate change and its impact on plant diseases. Current Science. 2013; 105(12):1685-1691.
22. Ghosh R, Nagavardhini A, Sengupta A, Mamta Sharma. Development of Loop-Mediated Isothermal Amplification (LAMP) assay for rapid detection of Fusarium oxysporum f. sp. ciceris wilt pathogen of chickpea. BMC Research Notes. 2015; 8:40.
23. Goswami K, Sharma R, Singh PK, Singh G. Micropropagation of seedless lemon (*Citrus limon* L. cv. *Kaghzi Kalan*) and assessment of genetic fidelity of micropropagated plants using RAPD markers. Physiology and Molecular Biology of Plants. 2013; 19(1):137-45.
24. Gupta PK, Roy JK, Prasad M. Single nucleotide polymorphisms: a new paradigm for molecular marker technology and DNA polymorphism detection with emphasis on their use in plants. Curr Sci. 2001; 80:524-535.
25. Gupta SK, Charpe A, Koul S. Development and validation of molecular markers linked to an Aegilops umbellulata-derived leaf rust-resistance gene. Genome. 2005; 48:823-830.
26. Gutherie PA, Magill CW, Frederiksen RA, Odvody GN. Random amplified polymorphic DNA markers: a system for identifying and differentiating isolates of Colletotrichum graminicola. Phytopathology. 1992; 82:832-825.
27. Hahn ML, Burrows PR, Gyanapragasam NC, Bridge J, Vines NJ, Write P. Molecular diversity in *Radopholus similis* population from Sri Lanka

detected by RAPD analysis. Fundamental Appl. Nematol. 1994; 17:275-281.

28. Helguera M, Khan IA, Dubcovsky J. Development of PCR markers for the wheat leaf rust resistance genc Lr47. Theor. Appl. Genet. 2000; 100:1137-1143. Doi: 10.1007/s001220051397.

29. Herrera-Foessel SA, Singh RP, Huerta-Espino J, William HM, Rosewarne G, Djurle A. Identification and mapping of Lr3 and a linked leaf rust resistance gene in durum wheat. Crop Sci. 2007; 47:1459-1466. Doi: 10.1007/s00122-012-1788-8.

30. Jinek M, Chylinski K, Fonfara I, Hauer M, Doudna JA, Charpentier E. A programmable dual-RNA-guided DNA endonuclease in adaptive bacterial immunity. Science. 2012; 337(6096):816-821.

31. Kaymak S, Boyraz N, Daniels J. Molecular markers to evaluate genetic diversity among Venturia inaequalis isolates obtained from apple plantations in Isparta Province. Turk J Agric For. 2016; 40:489-498.

32. Keshavarz K, Sijam K, Zainal Abidin MA, Habibuddin H, Nejat N, Nazerian E. Genetic diversity of Xanthomonas oryzae pv. Oryzae strains from rice fields in Malaysia. J Pl. Pathol. 2011; 93(3):719-724.

33. Khiyami MA, Almoammar H, Awad YM, Alghuthaym MA, Abd-Elsalam KA. Plant pathogen nanodiagnostic techniques: forthcoming changes? Biotechnology and Biotechnological Equipment. 2014; 28(5):775-785. doi.org/10.1080/13102818.

34. Liu S, Yu LX, Singh RP, Jin Y, Sorrells ME, Anderson JA. Diagnostic and co-dominant PCR markers for wheat stem rust resistance genes Sr25 and Sr26. Theor.Appl.Genet. 2014; 120:691-697. Doi: 10.1007/s00122-009- 1186-z.

35. Louws FJ, Fulkbright DW, Stephens CT, de Brujin FJ. Determination of genomic structure by rep-PCR fingerprinting to rapidly classify Xanthomonas campestris Pv. Vesicatoria. Phytopathology. 1995; 85:528-536.

36. Mago R, Bariana HS, Dundas IS, Spielmeyer W, Lawrence GL. Development of PCR markers for the selection of wheat stem rust resistance genes Sr24 and Sr26 in diverse wheat germplasm. Theor. Appl. Genet. 2005; 111:496-504. doi:10.1007/s00122-005-2039-z

37. Mago R, Verlin D, Zhang P, Bansal U, Bariana H, Jin Y. Development of wheat-Aegilops speltoides recombinants and simple PCR-based markers for Sr32 and a new stem rust resistance gene on the 2S-1

chromosome. Theor Appl. Genet. 2013; 126:2943-2955. Doi: 10.1007/s00122-013-2184-8.

38. Majer D, Mithern R, Lewis BG, Vos P, Oliver RP. The use of AFLP fingerprinting for the detection of genetic variation in fungi. Mycological Res. 1996; 100:1107-1111.

39. Mc Mullen MD, Louise R. The linkage of molecular markers to a gene controlling the symptom response in maize to maize dwarf mosaic virus. Mol. Plant-Microbe Interact. 1989; 2:309-314.

40. McCouch SR, Abenes ML, Angeles R, Khush GS, Tanksley SD. Molecular tagging of a recessive gene Xa-5, for resistance to bacterial blight of rice. Rice Genet. Newslett. 1991; 8:143-145.

41. McDonald BA, Miles J, Nelson R, Pettway RE. Genetic variability in nuclear DNA in field populations of Spongospora nodorum. Phytopathology. 1994; 84:250-255.

42. Messeguer R, Ganal M, de Vicente MC, Young ND, Bolkar H, Tanksley SD. High resolution RFLP map around the root knot nematode resistance gene (Mi) in tomato. Theor. Appl. Genet. 1991; 82:529-536.

43. Mew TN, Parco AR, Hittalmani S, Inukai T, Nelson R, Zeigher RS *et al.* Fine mapping of major genes for blast resistance in rice. Rice Genet. Newslett. 1994; 11:126-128.

44. Mohapatra T, Mukherjee AK, Srinivasan K, Sharma RP. Molecular markers for characterization of pathogen population and host genes. *In*: Molecular Approaches in Plant Disease Management. Eds. Shekawat GS, Singh BP and Singh RA. Malhotra Publishing House, New Delhi, India, 2002, 38-53.

45. Mundt CC. Landscape heterogeneity and disease spread: experimental approaches with a plant pathogen. Ecol. Appl. 2011; 21:321-328.

46. Neelam K, Brown-Guedira G, Huang L. Development and validation of a breeder-friendly KASPar marker for wheat leaf rust resistance locus Lr21. Mol. Breed. 2013; 31:233-237.doi:10.1007/s11032-012- 9773-0

47. Niessen L, Grafenhan T, Vogel RF. ATP citrate lyase 1 (acl1) gene-based loop-mediated amplification assay for the detection of the *Fusarium tricinctum* species complex in pure cultures and in cereal samples. Int J Food Microbiol. 2013; 158:171-185.

48. Niessen L, Vogel RF. Detection of *Fusarium graminearum* DNA using a loop-mediated isothermal amplification (LAMP) assay. Int J Food Microbiol. 2011; 140(2-3):183-191.

49. Nirmaladevi D, Venkataramana M, Rakesh K, Srivastava S, Uppalapati R, Gupta VK *et al*. Molecular phylogeny, pathogenicity and toxigenicity of *Fusarium oxysporum* f. sp. lycopersici. Scientific, Reports, 2016; 6:21367. DOI: 10.1038/srep21367.

50. Oerke EC. Crop losses to pests. J Agric. Sci. 2006; 144:31-43.

51. Patil S. Onion twister disease: Etiology, their characterization, epidemiology and integrated management. Ph.D. Thesis, Univ. Agric. Sci. Dharwad, Karnataka (India), 2013.

52. Periyannan S, Bansal U, Bariana H, Deal K, Luo MC, Dvorak J. Identification of a robust molecular marker for the detection of the stem rust resistance gene Sr45 in common wheat. Theor. Appl. Genet. 2014; 127:947-955.doi: 10.1007/s00122-014-2270-6.

53. Phan HTT, Ford R, Taylor PWJ. Population structure of *Ascochyta rabiei* in Australia based on STMS fingerprints. Fungal Diversity. 2003; 13:111-29.

54. Polzin KM, Lorenzen LL, Olson TC, Shoemaker RC. An unusual polymorphic locus useful for tagging Rps 1 resistance alleles in soybean. Theor. Appl. Genet. 1994; 89:226-232.

55. Qiu JW, Schürch AC, Yahiaoui N, Dong LL, Fan HJ, Zhang ZJ. Physical mapping and identification of a candidate for the leaf rust resistance gene Lr1 of wheat. Theor. Appl. Genet. 2007; 115:159-168. Doi: 10.1007/s00122-007-0551-z.

56. Raghu S. Studies on Chilli wilt complex disease. *Ph. D. Thesis*, Univ. Agric. Sci. Dharwad, Karnataka (India), 2007.

57. Rick CM, Forbes J. Association of an allozyme with nematode resistance. Tomato Genetics Cooperative Report. 1974; 24-25.

58. Ritter E, Debener T, Barone A, Salamini F, Gebhardt C. RFLP mapping on potato chromosomes of two genes controlling extreme resistance to potato virus X (PVX) Mol. Gen. Genet. 1991; 227:81-85.

59. Ronald PC, Albano B, Tabien R, Abenes L, Wu K, McCouch S *et al*. Genetic and physical analysis of the rice bacterial blight resistance locus, Xa-21. Mol. Gen. Genet. 1992; 236:113-120.

60. Rong Z, Yuan Y, Wang X, Zheng X. Rapid diagnosis of rice bakanae caused by Fusarium fujikuroi and F. proliferatum using loop-mediated isothermal amplification assays. Journal of Phytopathology, 2017. DOI: 10.1111/jph.12685.

61. Sarfatti M, Katan J, Fluhr R, Zamir D. An RFLP marker in tomato

linked to the Fusarium oxysporum resistance gene I_2. Theor. Appl. Genet. 1989; 78:755-759.

62. Schnick D, Rumpenhorst HJ, Burgermeister W. Differentiation of closely related Globodera pallid population by means of restriction fragment length polymorphism. J Phytopathol. 1990; 130:127-136.
63. Shankar P. Molecular mapping of blast resistance gene in rice (*Oryza sativa*). M.Sc. Thesis. Indian Agricultural Research Institute, New Delhi, 2002, 60.
64. Sharma TR, Tewari JP. Detection of genetic variation in Alternaria brassicae by RAPD fingerprints. J Plant Biochem Biotechnol. 1995; 4:105-107.
65. Singh BK. Molecular tagging of rice blast resistance gene using PCR based markers. M Sc Thesis. Indian Agricultural Research Institute, New Delhi, 2002, 53.
66. Singh BD, Singh AK. Marker-Assisted Plant Breeding: Principles and Practices, 2015. DOI: 10.1007/978-81-322-2316-0.
67. Singh S, Bowden RL. Molecular mapping of adult-plant race-specific leaf rust resistance gene Lr12 in bread wheat. Mol. Breed. 2011; 28:137-142. Doi: 10.1007/s11032-010-9467-4.
68. Singh R, Goutam U, Gupta RK, Pandey GC, Shoran J, Tiwari R. Allelic variations of functional markers for polyphenol oxidase (PPO) genes in Indian Wheat (*Triticum aestivum* L.) cultivars. J Genet. 2009; 88:325-329. doi: 10.1007/s12041-009-0047-5.
69. Sommerhalder RJ. Effect of hosts on competition among clones and evidence of differential selection between pathogenic and saprophytic phases in experimental populations of the wheat pathogen Phaeosphaeria nodorum. BMC Evol. Biol, 2011; 11:188.
70. Southern EM. Detection of specific sequences of DNA fragments separated by gel electrophoresis. J Mol. Biol. 1975; 98:503-517.
71. Srinivasan K, Gilardi G, Spadaro D, Gullino ML, Garibaldi A. Molecular characterization through IGS sequencing of formae speciales of Fusarium oxysporum pathogenic on lamb's lettuce. Phytopathol. Mediterr. 2010; 49:309-320.
72. Stepien L, Golka L, Chelkowski J. Leaf rust resistance genes of wheat: identification in cultivars and resistance sources. J Appl. Genet. 2003; 44:139-149.
73. Sujkowski LS, Goodwin SB, Dyer AT, Fry WE. Increased genotypic

diversity via migrationand possible occurance of sexual reproduction of Phytopthora infestans in Poland. Phytopathology. 1994; 84:201-207.

74. Terracciano I, Maccaferri M, Bassi F, Mantovani P, Sanguineti MC, Salvi S. Development of COS-SNP and HRM markers for high throughput and reliable haplotype-based detection of Lr14a in durum wheat (Triticum durum Desf.). Theor. Appl. Genet. 2013; 126:1077-1101. Doi: 10.1007/s00122-012- 2038-9.

75. Tsilo TJ, Jin Y, Anderson JA. Diagnostic microsatellite markers for the detection of stem rust resistance gene Sr36 in diverse genetic backgrounds of wheat. Crop Sci. 2008; 48:253-261. Doi: 10.2135/cropsci2007. 04.0204

76. Udupa SM, Weigand F, Saxena MC, Kahl G. Genotyping with RAPD and microsatellite markers resolves pathotype diversity in the ascochyta blight pathogen of chickpea. Theor Appl Genet. 1998; 97:299-307.

77. Valente MT, Desiderio F, Infantino A, Vale G, Abbruscato P, Aragona M. Genetic variability of *Fusarium fujikuroi* populations associated with bakanae of rice in Italy. Plant Pathol. 2017; 66:469-479.

78. Vander Beek JG, Verkerk R, Zabel P, Lindhout P. Mapping strategy for resistance genes in tomato based on RFLPs between cultivars: Cf 9 (resistance to *Cladosporium fulvum*) on the chromosome 1. Theor. Appl. Genet. 1992; 84:106-112.

79. Wang GL, Mackill DJ, Bonman M, McCouch SK, Champoux M, Nelson R. RFLP mapping of genes conferring complete and partial resistance to blast in a durably resistant rice cultivar. Genetics. 1994; 136:1421-1434.

80. Wang F, Wang C, Liu P, Lei C, Hao W, Gao Y *et al*. Enhanced rice blast resistance by CRISPR/Cas9-targeted mutagenesis of the ERF transcription factor gene OsERF922. PloS one. 2016; 11(4):e0154027.

81. Wang Y, Cheng X, Shan Q, Zhang Y, Liu J, Gao C *et al*. Simultaneous editing of three homoeoalleles in hexaploid bread wheat confers heritable resistance to powdery mildew. Nature Biotechnology. 2014; 32(9):947-951.

82. Xia JQ, Correl JC, Lee FN, Marchetti MA, Rhoads DD. DNA fingerprinting to examine microgeographic variation in the *Magnoporthe grisea* population in two rice fields in Arakansas. Phytopathology. 1993; 83:1029-1035.

83. Yli-Mattila T, Mironenko NV, Alekhina IA, Hannukkale A, Bulat SA.

Universally primed polymerase chain reaction analysis of *Fusarium avenaceum* isolated from wheat and barley in Finland. Agric. Food. Sci. Fin. 1997a; 6:25-36.

84. Yli-Mattila Y, Paavanen-Huhtala S, Bulat SA, Alekhina IA, Nirenberg HI. Molecular, morphological and phylogenetic analysis of *Fusarium avenaceum, F. arthrosporioides, F. tricinctum* species complex-a polyphasic approach. Myco. Res. 2002; 106: 655-669.

85. Yli-Mattila Y, Paavanen-Huhtala S, Hannukkala A, Alekhina IA, Hyvonen J, Bulat SA, RAPD-PCR, UP-PCR, rDNA RFLP, rDNA sequence and isozyme analysis of *Fusarium* isolates in Finland. Cereal Res. Comm. 1997b; 25:629-630.

86. Yoshimura S, Yoshimura A, Saito A, Kishimoto N, Kawase M, Yano M *et al*. RFLP analysis of introgressed chromosomal segments in three near-isogenic lines of rice bacterial blight resistance genes, Xa-1, Xa-3 and xa-4. Jpn. J Genet. 1992; 67:29-37.

87. Yu YG, Mackill DJ, Bonman JM, Tanksley SD. Tagging genes for blast resistance in rice via linkage to RFLP markers. Theor. Appl. Genet. 1991; 81:471-476.

88. Yu YG, Mackill DJ, Bonman JM, Tanksley SD. Tagging genes for blast resistance in rice via linkage to RFLP markers. Theor. Appl. Genet. 1991; 81:471-476.

89. Zheng KL, Zhuang JY, LU J, Qian HR, Lin HX. Identification of DNA marker linked to blast resistance genes in rice. Paper presented at the FAO/IAEA International Symposium on the use of induced mutations and molecular techniques for crop improvement Vienna, Austria, 1995, 19-23.

90. Zhu H, Zhai WX, Chen XL, Wang JL, Lei CL, Ling ZZ *et al*. Repeated sequence POR 6 and typing of Pyricularia oryzae isolates. Acta Mycologia Sinica. 1995; 14:218-225.

Chapter - 3
Exotic Spiralling Whitefly, *Aleurodicus dispersus* - Distribution & Management in India

Authors

Boopathi T.
ICAR-Research Complex for NEH Region, Mizoram Centre, Kolasib –796081, Mizoram, India

Sankari Meena K.
ICAR-National Rice Research Institute, Cuttack- 753006, Odisha, India

ArunKumar N
Department of Agricultural Microbiology, Tamil Nadu Agricultural University, Coimbatore, Tamil Nadu, India

Chapter - 3

Exotic Spiralling Whitefly, *Aleurodicus dispersus* - Distribution & Management in India

Boopathi T, Sankari Meena K and N Arunkumar

1. Introduction

The spiraling whitefly, *Aleurodicus dispersus* Russell (Hemiptera: Aleyrodidae), is an insect pest of many tropical and sub-tropical crops. The spiraling whitefly, native to the Caribbean region of Central America (Russell, 1965), is a highly polyphagous pest. In India, it was first recorded in 1993 at Thiruvananthapuram, Kerala on cassava (Palaniswami et al., 1995). The whitefly might have been introduced into India from neighboring countries like the Maldives (Muniappan, 1996), Sri Lanka (Ranjith et al., 1996) and Myanmar (Boopathi, 2008; Boopathi et al., 2014a) through infected planting materials. Since 1993 the species has spread throughout the southern and northeastern India, thereafter it has become an important pest of agricultural and horticultural crops (Boopathi et al., 2013; Boopathi et al., 2015a). Direct feeding damage was caused by the piercing and sucking of sap from foliage by immature and adult stages of whiteflies (Boopathi et al., 2012). Infestation of dense population causes premature leaf drop and production of copious amount of honeydew serves as a substrate for sooty mould growth (Akinlosotu et al., 1993; Boopathi et al., 2014b). Sooty mould blackens the leaves, decreases photosynthetic activity and vigour of the host plants (Kumashiro et al., 1983; Boopathi et al., 2014c; Boopathi et al., 2015b).

Aleurodicus dispersus is a polyphagous pest with an extensive host range including many crops and weed species (Srinivasa, 2000; Boopathi et al., 2012; Boopathi and Karuppuchamy, 2013). In India, the pest was first reported on tapioca by Palaniswami *et al.*, 1995. Ranjith *et al.*, 1996 enlisted the breeding hosts and feeding hosts of *A. dispersus* on coconut, banana, eggplant, okra, chillies, sapota, citrus, pepper, tomato, jack fruit, cashew nut, cocoa, pigeon pea, papaya, guava, cassia, fig, jasmine, leucinia, mango, holy basil, sweet basil, rose and castor. The present paper covers various aspects of the spiralling whitefly which will be useful to manage the pest.

2. Taxonomy

Aleurodicus dispersus was first described by Russell (1965) in Florida, USA and many Caribbean and Central American countries. *A. dispersus* belongs to the subfamily Aleurodicinae under family Aleyrodidae of the order Hemiptera. *A. dispersus* was characterized by distinctive compound and simple pores (Russell, 1965). *A. dispersus* is commonly called as spiralling whitefly because of the spiralling egg laying pattern. Mound and Halsey (1978) listed 1156 species in 126 genera of whiteflies (Aleyrodidae) in the catalogue of the whiteflies of the world and their host plants and natural enemies. Since then, several new genera and species have been described and others have been synonymized with previously described taxa.

3. Distribution

The spiralling whitefly, *Aleurodicus dispersus* Russell (Homoptera, Aleyrodidae) is native of the Caribbean region of Central America (Russell, 1965). It was reported to occur in North America, Central America, Caribbean islands, South America, Africa, Asia and several Pacific islands. In North America, it was first observed in 1957 in Southern Florida (Russell, 1965). Like most whiteflies, *A. dispersus* is of tropical and subtropical origin (Mound and Halsey, 1978). In India, *A. dispersus* was first reported in 1993 at Thiruvananthapuram on cassava and later from several other parts of Kerala, Tamil Nadu, Karnataka, Andhra Pradesh and Maharashtra. *A. dispersus* was distributed in Southern and Northeastern states of India and the intensity of damage ranged from 95.31 to 99.17 per cent in Tamil Nadu, Kerala and Karnataka. *A. dispersus* is reported for the first time in Meghalaya on rose (Boopathi, 2013). *A. dispersus* was found in all districts of Tamil Nadu and the intensity of damage ranged from 96.95 to 99.16 per cent in Coimbatore, Dharmapuri, Dindigul, Erode, Namakkal, Salem, Tiruchirappalli, Perambalur, Villupuram, Thiruvarur, Thanjavur, Madurai, Theni, Tirunelveli, Kanniyakumari, Krishnagiri and Tiruppur districts (Boopathi, 2013).

4. Morphology

Eggs were smooth, elliptical, whitish to yellow, translucent and deposited mainly on the axial surface of the leaves. Eggs were 0.2342 ± 0.0447 mm length and 0.0934 ± 0.0228 mm width (Boopathi, 2013). There were four distinct nymphal instars. The first instar nymphs called crawlers had functional walking legs and antennae. They were translucent, yellowish green, elliptical with a convex dorsum. The first instar nymphs were 0.1891 ± 0.0208 mm length and 0.0929 ± 0.0190 mm width. The second instar nymphs were oval, translucent and had many marginal fringes of wax covering the body of dorsum. The second instar nymphs were 0.3839 ± 0.0475 mm length and 0.2387 ± 0.0203 mm width. The third instar nymphs were oval with 0.6227 ± 0.0518 mm length and 0.4148 ± 0.0494 mm width. The fourth instar nymphs were 0.6667 ± 0.0804 mm length and 0.4317 ± 0.0334 mm width. The body was covered entirely with copious amount of white waxy materials. The length and width of pupae were 0.5099 ± 0.0468 mm and 0.3442 ± 0.0408, respectively. The pupae were differentiated from fourth instar by the presence of two red eye spots in the cephalic region. This stage is often called as red eyed nymphs. Adults emerged from the pupae through a 'T' shape exit slit on the dorsal surface of the pupae. Body length of adults was 1.2986 ± 0.1924 mm in male and 0.9633 ± 0.0188 mm in female with breadth of 0.2446 ± 0.0207 mm in male and 0.2973 ± 0.0401 mm in female (Boopathi, 2013).

5. Biology

Boopathi (2013) recorded that incubation period was ranging from 5.5 to 7.7 days on various host species. The mean duration of first instar nymph ranged from 2.3 to 6.5 days. The developmental duration of second instar nymphs ranged from 3.0 to 5.4 days. The duration of third instar nymph ranged from 3.2 to 6.1 days. The duration of fourth instar nymph was longer on *A. hispida* (8.0 days), *M. paradisiaca* (7.9 days) and *C. papaya* (7.4 days). The total developmental period of all nymphal instars ranged from 15.9 to 22.4 days. *M. paradisiaca* (22.4 days) recorded the longest period of total nymphal period (Boopathi, 2013). The shortest nymphal duration was recorded on *M. alba* (15.9 days). The pupal period ranged from 2.2 days to 3.4 days. The adult longevity ranged from 11.0 to 15.4 days. Fecundity showed greater variations among different host plants. *M. esculenta* (28.5 per female) recorded the maximum number of eggs laid per female. Percent adult emergence ranged from 88.4 to 98.2. Egg hatchability ranged from 91.9 to 98.7 per cent (Boopathi, 2013). Development from egg to adult occupied 30 days (Esguerra, 1987), 34 to 38 days at 20 to 39^{0}C (Waterhouse and Norris, 1989), 22 to 47 days (Wijesekera and Kudagamage, 1990) and 18 to 23 days (Palaniswami *et al.*, 1995) and 22.5 to 29.66 days (Geetha, 2000). Aiswariaya *et al.* (2007a) recorded the total development period of *A. dispersus* from 45.48 to 53.40 days in rainy season, 58.46 to 70.76 days in winter season and 35.81 to 41.83 days in summer season.

6. Ecology

Boopathi (2013) reported that *A. dispersus* population density was high during second fortnight of February 2012 (363.0 per 5 plants). *A. dispersus* population was comparatively higher from first fortnight of December 2011 (271.2 per 5 plants) till the second fortnight of May 2012 (233.8 per 5 plants) and then declined to low levels up to November 2012 and there was an increase in population from first fortnight December 2012 (235.8 per 5 plants) to second fortnight of February 2013 (276.4 per 5 plants). Correlation between abiotic factors and *A. dispersus* population showed that maximum temperature (r = 0.388*) and sun shine (r = 0.553**) recorded positive significant correlations (Boopathi, 2013). There was a negative significant correlation between evening relative humidity (r = -0.584**) and wind speed (r = -0.501**).

7. Host Plants

Aleurodicus dispersus is highly polyphagous and it is known to attack as many as 491 plants in different countries in which the pest was introduced. In India, the pest was first reported in 1993 on tapioca by Palaniswami *et al.* (1995). *A. dispersus* was recorded on the plants numbering 25 (David and Regu, 1995), 70 (Prathapan, 1996), 22 (Ranjith *et al.*, 1996), 53 (Asia Mariam, 1999), 27 (Gajendra Babu and David, 1999), 45 (Mani and Krishnamoorthy, 1999a), 94 (Muralikrishna, 1999), 128 (Geetha, 2000), 27 (Ramani, 2000), 68 (Srinivasa, 2000), 100 plants belonging to 38 genera and 27 families (Charati *et al.*, 2003) and 99 (Aiswariaya *et al.*, 2007a).

The incidence of *A. dispersus* was reported on various plant species from Kerala, Tamil Nadu and Karnataka. The pest was collected from wild tapioca and wild rubber in September 1994 and also from *Anacardium occidentale* L., *Manihot utlisima* Crantz; *Solanum* spp., and *Psidium guajava* L. from Marayapuram in Kanyakumari district of Tamil Nadu during 1995 (David and Regu, 1995). Palaniswarni *et al.* (1995) reported the incidence on cassava, annona, banana, okra, egg plant, cassia, citrus, chillies, coconut, fig, guava, jasmine, leucinia, mango, *Ocimum sanctum* L., *Ocimum basilicum* L., rose and sapota.

Ranjith (1998) enlisted the breeding hosts and feeding hosts of *A. dispersus*. *A. dispersus* recorded on coconut, banana, brinjal, lady's finger, chillies, sapota, citrus, pepper, tomato, jack, cashew, cocoa, pigeon pea, papaya, guava and castor. Among the ornamentals, its infestation was reported on rose, balsam, poinsettia, dahlia, heliconia, canna, acalypha and polyalthia. An ornamental shrub, *Dombeya spectabilis* Boj, medicinal plant, *Solanum trilobatum* L. and groundnut were recorded as new hosts for *A. dispersus* during field surveys in 1997-98 in Tamil Nadu (Geetha *et al.*, 1998 and 1999). Asia Mariam (1999) reported the plant species belonging to Euphorbiaceae, Solanaceae and Fabaceae were more susceptible to *A. dispersus* attack. Mani and Krishnamoorthy (1999c) observed the attack of *A. dispersus* on more than 45 species including fruit trees, ornamentals, vegetables, shade trees and other crops such as tapioca, cashew, mulberry and cotton in penninsular India. Boopathi (2013) reported that a total of 153 host plants belonging to 54 families were highly preferred by *A. dispersus* in India. Thirty nine weed species belonging to 15 families were identified as alternate hosts species including *Parthenium hysterophorus* L. in India. Totally 78 host plants were recorded for the first time at different spatial levels, 49 hosts were not recorded earlier in any country, 7 hosts were recorded for the first time in India and 22 hosts were recorded in Tamil Nadu for the first time (Boopathi, 2013).

8. Damage

8.1 Direct Damage

Nymphs and adults of *A. dispersus* congregated and infested heavily the lower surface of leaves of all the varieties of plants. They suck sap from their host plants and caused premature leaf drop (Esguerra, 1987). Infestation by *A. dispersus* caused extensive desapping (Akinlosotu *et al*., 1993) which led to chlorosis and early dehiscence of leaves (M' Boob and van Oers, 1994). Yellow speckling, crinkling and curling of the leaves was noted when the infestation by *A. dispersus* was severe (Palaniswami *et al*., 1995). Damage by *A. dispersus* in young plants, would lead to reduced growth. Tree species may escape death, but would have reduced vigour considerably (Ranjith, 1998). Wen *et al*. (1995) estimated that four month long infestation could result in 80% fruit yield loss in guava by *A. dispersus*.

Considering the wide spread and severe incidence of *A. dispersus* on a variety of plants in Kerala and Tamil Nadu, it is likely that this species may soon pose a threat to cultivation of economically important field and plantation crops in India (David and Regu, 1995). The occurrence and spread of this insect was alarming, attack by *A. dispersus* on rubber was likely to cause unseasonal leaf fall and consequent yield reduction by way of reduced latex flow. Production of banana reduced due to the attack of *A. dispersus* (Ranjith, 1998). *A. dispersus* had a dramatic negative effect on the agricultural industry, causing a significant amount of monetary loss to these enterprises (Alam *et al*., 1998). *Aleurodicus dispersus* initially released at the rate of 100, 250 and 500 per plot (5 x 4 m) on cassava showed significant differences in foliage weight and yield loss (Boopathi, 2013). Both foliage weight and yield of cassava infested with *A. dispersus* had significant differences when compared to uninfested plants. Foliage biomass per plant was 2.264, 2.015 and 1.939 kg per plant in plants released with 100, 250 and 500 *A. dispersus* per plot, whereas it was 2.939 kg per plant in uninfested plants. Average yield was 3.165 kg per plant from uninfested plants, whereas it was 2.114 kg per plant in plants infested with 100 *A. dispersus* per plot, 1.894 kg per plant in plants infested with 250 *A. dispersus* per plot and 1.201 kg per plant in plants infested with 500 *A. dispersus* per plot (Boopathi, 2013).

8.2 Indirect Damage

Some of the indirect damage caused by *A. dispersus* are (i) honeydew excretion and sooty mould development, (ii) secretion of waxy flocculent materials and covered in the leaves, (iii) act as vector for diseases transmission and (iv) nuisance to public. Much sticky honeydew excreted by *A. dispersus* served as a substrate for dense growths of sooty moulds which interfered with photosynthesis (Byrne *et al.,* 1990; Kajita and Alam, 1996). *A. dispersus* also transmitted virus diseases such as cassava brown streak virus (CBSV) (Mware *et al.*, 2009).

9. Natural Enemies

A total of 62 natural enemies, including 8 parasitoids, 53 predators and a fungal pathogen, are known to attack the whitefly in different countries (Mani and Krishnamoorthy, 1999a and 1999b; Mani and Krishnamoorthy, 2002).

9.1 Parasitoids

Gerling (1990) provided a short key to parasitoids of whiteflies. Seven parasitoids all belonging to Hymenoptera representing two families a parasite under Acarina have been recorded on *A. dispersus* (Table 1). *Encarsia guadeloupae* Viggiani and *Encarsia* sp nr *meritoria* Gahan were found as efficient parasitoids against *A. dispersus* in Tamil Nadu and Karnataka. The parasitism by *Encarsia* spp. was high during second fortnight of March 2012 and first fortnight of November 2012.

Table 1: Geographical records of parasitoids of *Aleurodicus dispersus* Russell

S. No.	Species	Family and order	Locatios
1.	*Aleuroctonus vittatus* (Dozier)	Aphelinidae (Hymenoptera)	USA, Cayman Islands, Nevis, Puerto Rico

2.	*Encarsia dispersa* Polaszek	Aphelinidae (Hymenoptera)	USA, Belize, Brazil, Trinidad, Venezuela, Benin, Taiwan (intro), India, Malaysia, Maldives, Kiribati, Philippines, Guam, Fiji, Hawaii
3.	*Encarsia guadeloupae* Viggiani	Aphelinidae (Hymenoptera)	Guam, Taiwan, USA, Guadeloupe, Trinidad, Canary Islands, Benin, Philippines, Thailand, Indonesia, Micronesia, Nevis, Papua New Guinea, Hawaii and India
4.	*Encarsia (?) haitiensis* Dozier	Aphelinidae (Hymenoptera)	Trinidad, Hawaii, Cook Is, American Samoa, Fiji, Torres Strait, Pohnpei, Guam, Palau, Taiwan, India, Ghana, Togo, Benin, Ibadan, Nigeria
5.	*E. (?) hispida*	Aphelinidae (Hymenoptera)	Canary Is
6.	*Encarsia lahorensis* (Howard)	Aphelinidae (Hymenoptera)	USA, Egypt, Greece, Iran, Israel, Italy, China, Taiwan, India and Pakistan
7.	*Encarsia nigricephala* Dozier	Aphelinidae (Hymenoptera)	USA, Barbados, Brazil, Colombia, Cuba, Ecuador, Grenada, Guadeloupe, Guatemala, Honduras, Jamaica, Martinique, Mexico, Puerto Rico, Venezuela, Reunion Islands, Micronesia, Florida and French Polynesia.
8.	*Encarsia sophia* (Girault and Dodd)	Aphelinidae (Hymenoptera)	USA, Honduras, Martinique, Iran, Italy, Morocco, Spain, Cape Verde Islands, Ivory Coast, Kenya, Niger, Sierra Leon, Somalia, Zimbabwe, China, Hong Kong, Japan, Taiwan, India, Pakistan, Thailand, Australia, Indonesia and Hawaii
9.	*Encarsia* sp.	Aphelinidae (Hymenoptera)	Guam, Hawaii, West Indies
10.	*Encarsia* sp nr *meritoria* Gahan	Aphelinidae (Hymenoptera)	India
11.	*Encarsia tabacivora* Viggiani	Aphelinidae (Hymenoptera)	USA, Brazil, Costa Rica, Honduras, Jamaica, Mexico, Puerto Rico.
12.	*Encarsia transvena* (Timberlake)	Aphelinidae (Hymenoptera)	Sri Lanka

13.	*Euderomphale vittata*	Aphelinidae (Hymenoptera)	Puerto Rico, Jamaica, Florida, Cayman Is, Sri Lanka
14.	*Encarsiella noyesi* Hayat	Aphelinidae (Hymenoptera)	Costa Rica
15.	*Encarsiella aleurodici* (Grlt.)	Aphelinidae (Hymenoptera)	Costa Rica
16.	*Leptus* sp.	Erythraeiclae (Acari)	India

9.2 Predators

A total of 49 species of predators belonging to 10 families representing 4 insect orders have been reported (Table 2). In the native home of West Indies, three species of coccinellids were found to be important predators on *A. dispersus* (Waterhouse and Norris, 1989). Boopathi (2013) reported that totally 28 species of predators including coccinellids, chrysopids, drosophilids, lycaenid, mantid and oxyopid were recorded. Among them, *Cybocephalus* spp., *Axinoscymnus puttarudriahi* Kapur, *Mallada astur* (Banks), *Acletoxenus indicus* Malloch, *Mallada boninensis* Okamoto and *Cryptolaemus montrouzieri* Muls were found to be potential predators. *Cybocephalus* spp., *A. puttarudriahi*, *M. astur* and coccinellid populations were more during first fortnight of July 2012 (12.40 per 5 plants), second fortnight of August 2012 (6.80 per 5 plants), first fortnight of September 2012 (13.40 per 5 plants) and first fortnight of February 2013 (4.60 per 5 plants).

Table 2: Geographical records of predators of *Aleurodicus dispersus* Russell

S. No.	Predator species	Family and order	Location
1.	*Anegleis cardoni*	Coccinelliclae (Coleoptera)	India
2.	*A. perrotteti*	Coccinelliclae (Coleoptera)	India
3.	*Axinoscymnus nigripennis*	Coccinelliclae (Coleoptera)	Taiwan
4.	*A. puttarudriahi*	Coccinelliclae (Coleoptera)	India, Sri Lanka
5.	*Axinoscymnus* sp.	Coccinelliclae (Coleoptera)	Indonesia
6.	*Bothrocalvia pupillata* (Swar.)	Coccinellidae, Coleoptera	Hawaii

7.	*Cheilomenes sexmaculatas*	Coccinelliclae (Coleoptera)	Taiwan, Indonesia, India
8.	*Chilocorus nigrita* (Fab.)	Coccinelliclae (Coleoptera)	India
9.	*Coelophora inaequalis*	Coccinelliclae (Coleoptera)	Indonesia
10.	*C. pupillata*	Coccinelliclae (Coleoptera)	Hawaii
11.	*Cryptolaemus montrouzieri*	Coccinelliclae (Coleoptera)	Hawaii, Taiwan, India
12.	*Curinus coeruleus*	Coccinelliclae (Coleoptera)	Hawaii, Philippines, India
13.	*Delphastus catalinae*	Coccinelliclae (Coleoptera)	Trinidad, Hawaii, American Samoa
14.	*Delphastus pusillas* (Leconte)	Coccinellidae, Coleoptera	Hawaii
15.	*Harmonia sedecimnotata*	Coccinelliclae (Coleoptera)	Indonesia
16.	*Horniolus* sp.	Coccinelliclae (Coleoptera)	India
17.	*Jauravia dorsalis*	Coccinelliclae (Coleoptera)	India
18.	*J. pallidula*	Coccinelliclae (Coleoptera)	India
19.	*Jauravia* sp.	Coccinelliclae (Coleoptera)	India
20.	*Keiscymnus* sp.	Coccinellidae, Coleoptera	India
21.	*Leis axyridis*	Coccinelliclae (Coleoptera)	Taiwan
22.	*Lemnia biplagiata*	Coccinelliclae (Coleoptera)	Taiwan
23.	*L. swinhoei*	Coccinelliclae (Coleoptera)	Taiwan
24.	*Megalocaria fijiensis*	Coccinelliclae (Coleoptera)	Fiji
25.	*Nephaspis bicolor*	Coccinelliclae (Coleoptera)	Trinidad, Hawaii, American Samoa, Fiji
26.	*Nephaspis namolica*	Coccinellidae, Coleoptera	Trinidad and Columbia
27.	*N. oculatus*	Coccinelliclae (Coleoptera)	Honduras, Trinidad, West Indies, Florida, Hawaii, American Samoa, Fiji, Guam
28.	*Nephaspis* sp.	Coccinellidae, Coleoptera	W. Africa

29.	*Nephus regularis*	Coccinelliclae (Coleoptera)	India
30.	*N. roepkei*	Coccinelliclae (Coleoptera)	Guam
31.	*Olla unigrum* var. *plagiata*	Coccinelliclae (Coleoptera)	Hawaii
32.	*O. v-nigrum*	Coccinelliclae (Coleoptera)	Hawaii
33.	*Platynaspidius maculosus*	Coccinelliclae (Coleoptera)	Taiwan
34.	*Propylea japonica*	Coccinelliclae (Coleoptera)	Taiwan
35.	*Pseudaspidimerus flaviceps*	Coccinelliclae (Coleoptera)	India
36.	*P. trinotatus*	Coccinelliclae (Coleoptera)	India
37.	*Pseudoscymnus* sp.	Coccinelliclae (Coleoptera)	India
38.	*Rodolia amabilis*	Coccinelliclae (Coleoptera)	India
39.	*R. breviuscula*	Coccinelliclae (Coleoptera)	India
40.	*R. cardinalis*	Coccinelliclae (Coleoptera)	Taiwan
41.	*R. fumida*	Coccinelliclae (Coleoptera)	India
42.	*R.pumila*	Coccinelliclae (Coleoptera)	Taiwan
43.	*Saula japonia*	Coccinelliclae (Coleoptera)	Taiwan
44.	*Scymnus coccivora*	Coccinelliclae (Coleoptera)	India
45.	*S. latemaculatus*	Coccinelliclae (Coleoptera)	India
46.	*S. nubilus*	Coccinelliclae (Coleoptera)	India
47.	*S. posticalis*	Coccinelliclae (Coleoptera)	India
48.	*S. saciformis*	Coccinelliclae (Coleoptera)	India
49.	*Scymnus* sp.	Coccinelliclae (Coleoptera)	Taiwan, India, Canary Is
50.	*Serangiella* sp.	Coccinelliclae (Coleoptera)	Fiji
51.	*Serangium maculigerum*	Coccinelliclae (Coleoptera)	Hawaii
52.	*S. parcesetosum*	Coccinelliclae (Coleoptera)	India
53.	*Serangium* sp.	Coccinelliclae (Coleoptera)	Taiwan
54.	*Cybocephalus*	Cybocephalidae	Taiwan

	taiwanensis	(Coleoptera)	
55.	*Cybocephalus* spp.	Cybocephalidae (Coleoptera)	Indonesia, India
56.	*Triommata coccidivora*	(Diptera)	India
57.	*Leucopis* sp.	Chamaemyiidae (Diptera)	India
58.	*Acletoxenus indicus*	Drosophilidae (Diptera)	India
59.	*Acletoxenus* sp.	Drosophilidae (Diptera)	Taiwan
60.	*Allograpta obliqua*	Drosophilidae (Diptera)	Hawaii
61.	*Paragus serratus*	Drosophilidae (Diptera)	Taiwan, Indonesia
62.	*Baccha* sp.	Syrphidae (Diptera)	Darwin
63.	*Paragus serratus* (Fab.)	Syrphidae, Diptera	Indonesia
64.	*Alloqraphata oblique* (say)	Syrphidae, Diptera	Hawaii
65.	*Iridomyrmex anceps*	Formicidae (Hymenoptera)	Indonesia
66.	*Oecophylla smaragdina*	Formicidae (Hymenoptera)	India
67.	*Solenopsis geminata*	Formicidae (Hymenoptera)	India
68.	*Spalgis epeus*	Lycaenidae (Lepidoptera)	India
69.	*Apertochrysa* sp.	Chrysopidae (Neuroptera)	India
70.	*Chrysopa* sp.	Chrysopidae (Neuroptera)	Fiji, Indonesia
71.	*Chrysoperla carnea*	Chrysopidae (Neuroptera)	India
72.	*C. comanche*	Chrysopidae (Neuroptera)	Hawaii
73.	*Mallada astur*	Chrysopidae (Neuroptera)	India
74.	*M. boninensis*	Chrysopidae (Neuroptera)	Taiwan, India
75.	*Nobilinus* sp.	Chrysopidae (Neuroptera)	India
76.	*Sympherobius barberi*	Chrysopidae (Neuroptera)	Hawaii
77.	*Hemerobius* sp.	Hemerobiidae (Neuroptera)	India
78.	*Notiobiella viridinervis*	Hemerobiidae (Neuroptera)	India
79.	*Orius* sp.	Anthacoridae (Hemiptera)	Taiwan
80.	*Oxyopus* sp.	Oxypidae, Acari	India

81.	*Bufo gargarizans*	Toad	Taiwan
82.	*Passer domesticus*	Aves	India
83.	*Arachnothera longirostris*	Aves	India
84.	*Saxicola caprata*	Aves	India
85.	Sunbirds	Aves	India
86.	*Orthotomus sutorius*	Aves	India

9.3 Entomopathogens

A total of 5 entomopathogens reported from different countries especially in India and Taiwan (Table 3). Wraight *et al.* (1998 and 2000) observed equal pathogenicity of *P. fumosoroseus* and *B. bassiana* against nymphs of *B. argentifolii* under laboratory conditions. *Paecilomyces farinosis* (Holms.) was known to infect the nymphs of spiralling whitefly in Bangalore (Mani et al., 2001). The nymphs were found infected with *Verticillium lecanii* Zimm. at Dharwad (Mallappanavar, 2000). *Fusarium semitectum* and V.lecanii were known ti cayse 65-70% and 80-90% moratlity of spiralling whitefly (Aiswariya et al., 2007b).

Table 3: Geographical records of entomopathogens of *Aleurodicus dispersus* Russell

S. No.	Entomopathogens	Class	Location
1.	*Pacilomyces farinosus* (Brown and Smith)	Moniliales, Deutromycetes	India
2.	Verticillium *lecanii* Zimm. Viegas	Moniliales, Deutromycetes	India
3.	*Acremonium* sp.	Moniliaceae	Taiwan
4.	*Aschersonia aleyrodis*	Nectrioidaceae	Taiwan

10. Management

Management of *A. dispersus* became all the more difficult because of the multitude of host plants that grow wild in nature support the build-up of the pests. The primary goal of integrated pest management (IPM) is to optimize pest control in an economically and ecologically sound way. IPM involves the integration of resistant varities, cultural, physical, mechanical, biological, and chemical practices to grow crops with minimal use of pesticides. Monitoring, sampling, and record keeping are used to determine when control options are needed to keep pests below an economically damaging threshold. Pest management, not eradication, is the goal of IPM.

10.1 Cultural Control

Use of clean planting material delays the appearance of the spiralling whitefly population. Pruning the heavily infested trees and shrubs was recommended to reduce the *A. dispersus* incidence. Subsequent to the pruning the population rapidly increased within 4-5 months on guava (Geetha, 2000).

10.2 Physical Control

Physical control of *A. dispersus* reported by Banjo and Latunde-Dada (1999), Banjo *et al.* (2003) and Banjo *et al.* (2004). The population of *A. dispersus* during wetter season declined drastically as sporadic rainfall washed off the eggs and nymphs from the host leaves. Banjo *et al.* (2004) therefore concluded that intensive spraying of the underside of leaves with water also reduced the population of *A. dispersus*. Light trap was more appropriate tool for monitoring. Fluorescent light smeared with castor oil attracted and trapped large number of adults (Asia Marium, 1999). Boopathi (2013) reported that yellow sticky traps recorded the highest attraction of *A. dispersus* adults and the sticky traps placed below crop canopy level in cassava and the sticky traps placed at crop canopy level in brinjal attracted more whiteflies.

10.3 Botanical Control

Kavitha Kirubavathy *et al.* (1999) observed the neem products such as 2% and 3'% neem oil (NO), 2% neem seed kernel extract (NSKE) and neem oil (2%) + NSKE (3%) were found to be effective in suppressing the nymphal and adult whitefly population. Tobacco extract (4%) was found effective in minimizing the *A. dispersus* (Muralikrishna, 1999). Spraying of neem oil (2%) reduces the whitefly population (Ranjith *et al.*, 1996; Asia Mariam, 1999 and Geetha, 2000). Geetha (2000) observed that spraying of TNAU neem formulation noticed maximum adult mortality (92.66%) which was on par with phosalone (92.65%). A dilute aqueous solution of a dish washing detergent suggested as helpful for the management of *A. dispersus* (Waterhouse and Norris, 1989). FORS was found to be effective against *B. tabaci* (Venugopal Rao *et al.*, 1990). Natarajan *et al.* (1991) tested the effectiveness of FORS along with chemicals against *B. tabaci*. FORS treated plot has harboured minimum population of 15.3 nymphs per leaf followed by neem oil (18.0 per leaf) as compared to maximum population (60.6) in monocrotophos. FORS along with phosalone also resulted in less whitefly population. Per cent egg mortality and nymphal mortality were maximum with three applications of 5% soap solution (Ranjith *et al.*, 1997). Spraying of FORS (4%) and detergent soap solution (5%) reduced the whitefly population (Ranjith *et al.*, 1996; Asia Mariam, 1999 and Geetha, 2000).

Boopathi (2013) recorded that botanicals, FORS and organic salt were found to cause substantial reduction. NSKE 5% was found to be the most effective in reducing the population of *A. dispersus* on cassava (more than 85 per cent) at 10 DAT, which was followed by neem oil 3% (78.8 per cent). Similarly, NSKE 5% recorded the highest mortality of *A. dispersus* (more than 77 per cent) on brinjal (Boopathi, 2013).

10.4 Biological Control

Since, *A. dispersus* is not easily amenable to chemical control, hence biological control would be the tactic of prime choice to manage this pest. Several attempts were made to search natural enemies in its native home lands and to introduce the natural enemies to the places where this insect was introduced. Synthetic insecticides did not adequately control this whitefly since the nymphs were covered with heavy waxy flocculent materials. Only the adults were susceptible to the insecticidal applications. Application of insecticides would temporarily reduce the whitefly abundance (Waterhouse and Norris, 1989). Even if the whitefly is controlled on some plants, there is heavy migration from roadside trees to the cultivated crops. Therefore, alternate methods such as biological control could help in the suppression of *A. dispersus*.

10.4.1 Parasitoids

In Malaysia, the populations of *A. dispersus* were very low mainly due to higher parasitism (> 80 %) by *E. guadeloupae* (Kajita *et al.*, 1991). *E. haitiensis* and *Encarsia* sp. were known to cause high parasitism on *A. dispersus* in the countries to which these two parasitoids were accidentally or deliberately introduced (Waterhouse and Norris, 1989). *A. dispersus* is naturally controlled in the Canaries by *Encarsia hispida* (De Santis) and also by the introduced *E. guadeloupae* (Nijhof *et al.*, 2000; Hernandez-Suarez *et al.*, 2003). Srinivasa *et al.* (1999) reported the occurrence of *E. haitiensis* in Bengaluru, Karnataka and also showed that the per cent parasitism was influenced by the host plant. The per cent parasitism ranged from nil to 38.88 on different host plants being highest on *C. siamea* (Fabaceae). Ramani (2000) reported both *E. haitiensis* and *E. guadeloupe* from Lakshadweep Islands (India).

The aphelenid parasitoids, *Encarsia haitiensis* Dozier and *Encarsia guadeloupae* Viggiani had given excellent control of *A. dispersus* in several countries (Waterhouse and Norris, 1989; D'Almeida *et al.*, 1998; Mani and Krishnamoorthy, 2000). Parasitoid, *E. guadalopae* was found to be the most effective in the reduction of *A. dispersus* population both after 60 days of first (82.6 %) and second releases (96.0 %) followed by *E.* sp nr *meritoria* in both the releases (79.6 and 92.2 %, respectively) in cassava (Boopathi, 2013). Sixty days after second release, all the three species of parasitoids were found to be more effective causing more than 90% reduction in *A. dispersus* population. In brinjal also, the release of *E. guadalopae* was found to be the effective parasitoid in reduction of *A. dispersus* population which was followed by the release of *E.* sp nr *meritoria*.

10.4.2 Predators

According to Paulson and Kumashiro (1985), twelve predators were found attacking the *A. dispersus* in Hawaii. Of them, *Delphastus pusillus* Leconte, *Nephaspis oculatus* (=*N. amnicola*) Wingo and *N. bicolor* Gordon showed strong preference for whiteflies. The other general predators like reduviids, spiders and lizards and birds were never able to control the *A. dispersus* in Hawaii. Native predators did not have any significant impact on the population of *A. dispersus* in Fiji (Kumar *et al.,* 1987), Guam (Nechols, 1982 and 1983) and India (Mani *et al.,* 2000). Eight species of predators were found attacking the *A. dispersus* on guava, but they failed to suppress the outbreaks of *A. dispersus* in Indonesia (Kajita *et al.*, 1991). In Sri Lanka, *A. puttarudriahi* Kapur and Munshi appeared to be a potential predator of the *A. dispersus*, but its impact in reducing the whitefly was not assessed (Wijesekera and Kudagamage, 1990). In India, predators numbering 22 (Mani *et al.*, 2000) and 15 (Geetha, 2000) were known to attack *A. dispersus*. Among them, *Cryptolaemus montrouzieri* Muls. *Mallada astur* (Banks), *Cybocephalus* spp. and *A. puttarudriahi* appeared in more numbers at times (Mani and Krishnamoorthy, 1997a and 1997b; 1999a and 1999b). *A. puttarudriahi* and *Cybocephalus* spp. did not prefer to feed the nymphs parasitised by *E. quadeloupae* (PDBC, 2000).

The two native predators, *A. puttarudiahi* and *Cybocephalus* spp. (Coleoptera: Nitidulidae) were able to discriminate between the parasitized and healthy larvae and pupae (Ramani, 2000). The predator *C. montruozieri* was found throughout the year. *A. puttarudriahi* was found only during July second fortnight to October first fortnight. *Chrysoperla zastrowi sillemi* (Esben - Petersen) was found during October second fortnight to December first fortnight. However, the density of natural enemies was found to be very low (Aiswariaya *et al.*, 2007b). The native predators were incapable of preventing the outbreak of whitefly on guava in Indonesia (Kajita *et al.*, 1991). Boopathi (2013) reported that inundative releases of *M. astur* was found to be the most efficient predator in reducing the population of *A. dispersus* after 60 days of first (84.2%) and second releases (97.0 %) followed by *Cybocephalus* spp. (79.8 %). In brinjal, *M. astur* was found to be the efficient predator in reducing the population of *A. dispersus* in both first (78.3 per cent) and second releases (98.4 %) at 30 days after release (DAR) followed by *Cybocephalus* spp. in both first (72.5 %) and second releases (97.0 %).

10.4.3 Entomopathogens

Entomopathogenic fungi, *P. fumosoroseus* recorded significantly maximum mortality in both first (73.70 %) and second sprays (79.96 %) at 15 days after spray (DAS) in cassava. In brinjal, *P. fumosoroseus* recorded maximum mortality in both first (80.21 %) and second sprays (82.43 %) at 15 DAS (Boopathi, 2013). Aiswariaya *et al.* (2007b) reported that the *F. semitectum* @ 4.2 x 10^9 spores/ml recording a per cent mortality of 75.21 (nymphs) and 64.40 per cent (adults), while *V. lecanii* 3.6 x 10^9 spores/ml recording a mortality of 89.97 (nymphs) and 79.90 per cent (adults) at 6 days after spray. *F. semitectum* 6.2 x 10^{15} + *L. lecanii* 4.6 x 10^{14} + Triazophos (0.03%) and *L. lecanii* 4.6 x 10^{14} + Triazophos (0.03%) recorded significantly maximum mortality under field conditions (Aiswariaya *et al.*, 2007b). The fungus *Lecanicillium lecanii* (Zimmerm.) Zare and Gams was reported to be effective against *A. dispersus* (Mallappanavar, 2000) and the infectivity of *Fusarium semitectum* Berk. and Ravenel on nymphs of *A. dispersus* was also reported (Mikunthan, 2004). Mikunthan (2004) reported the higher efficacy of the fungus, when it was used in combination with insecticides suppressed the whitefly population compared to the fungus alone. The principle behind the use of sub lethal dose of insecticide, in combination with mycopathogen is to weaken the insects by the lower dose of the toxicant that ultimately predispose to fungal infection.

10.5 Chemical Control

Megir-Gumbek (1987) tested nine insecticides against *A. dispersus* on guava and observed significant reduction in damage and infestation. On low growing vegetables, spraying with either malathion or diazinon with spreader or sticker can control whitefly. While spraying, the nozzle should be directed to the underside of leaves where the whitefly is most often located (Esguerra, 1987). The contact insecticides temporarily reduced whitefly abundance and systemic insecticides would be even more effective (Waterhouse and Norris, 1989). Spraying of buprofezin and dimethoate at high volume was effective for controlling this pest on guava (Wijesekera and Kudagamage, 1990). Minimum whitefly population of *B. tabaci* was recorded in triazophos treatment followed by amitraz and phosalone on cotton (Natarajan *et al.*, 1991). Wen *et al.* (1995) recommended bifenthrinate 2.8% EC for best egg control and methomyl for nymphal and adult mortality. Contact insecticides (Malathion at 1ml/litre of water and carbaryl at 2g/l of water) were also found effective against young nymphs (Ragumoorthy and Kempraj, 1996).

Use of chemicals like thiometon, dimethoate and phospharnidon could help to reduce the pest population to some extent. Chemicals like triazophos (0.04 %) and phosalone (0.05%) recommended for other whiteflies can help to suppress the *A. dispersus* population only to some extent. Moreover, migration habit of *A. dispersus* from one field to other, made chemical control more difficult and expensive (Mani and Krishnamoorthy, 1997a). A mixture of buprofezin and thiocylam showed the best effect in reducing the *A. dispersus* population in Costa Rico (Laprade and Gerdas, 1998). Dichlorvos 0.08% was found toxic to various stages of *A. dispersus* (Asia Mariam, 1999). Triazophos 0.08% and phosalone 0.07% were equally effective against *A. dispersus* (Geetha, 2000).

Boopathi (2013) reported that Application of acephate (99.24 %) and triazophos (98.46 %) recorded the highest reduction on the *A. dispersus* population in cassava at 10 DAT which were on par with each other. Application of acephate (98.07 %), triazophos (97.68 %), acetamiprid and thiocloprid (96.39 %) recorded the highest reduction on the *A. dispersus* population in brinjal at 10 DAT.

11. References

1. Aiswariaya KK, Manjunatha M, Naik M. Biology and host range of spiralling whitefly. Karnataka Journal of Agricultural Sciences. 2007a; 20:149-152.

2. Aiswariaya KK, Manjunatha M, Naik M. Evaluation of fungi *Fusarium semitectum* Berk and Ravenel and *Verticillium lecanii* (Zimm.) Viegas against spiraling whitefly Aleurodicus dispersus Russell on guava. Karnataka Journal of Agricultural Sciences. 2007b; 20:283-287.

3. Akinlosotu TA, Jackai LEN, Nitonifor NN, Hassan AT, Agyakwa GW, Odfbiyi JA *et al.* Spiralling whitefly, *Aleurodicus dispersus* in Nigeria.

FAO Plant Protection Bulletin. 1993; 41:127-129.

4. Alam S, Islam MN, Alam MZ, Islam MS. Effectiveness of three insecticides for the control of the spiralling whitefly (Aleurodìcus dispersili Rüssel) of guava. Bangladesh Journal of Entomology. 1998; 8:53-58.

5. Asia Mariam. Biology and Management of spiralling whitefly *Aleurodicus dispersus* (Russell) (Homoptera: Aleyrodidae) on mulberry. *M.Sc. (Ag.) Thesis*, Tamil Nadu Agricultural University, Coimbatore, 1999, 88.

6. Banjo AD, Latunde-Dada IL. An assessment of host plant preference of the spiralling whitefly (*Aleurodicus dispersus*) in Ago-Iwoye, Nigeria. Journal of Crop Research. 1999; 17(3):390-394.

7. Banjo AD, Hassan AT, Jackal LEN, Dixon AGO, Ekanayake IJ. Developmental and Behavioural study of spiralling whitefly (*Aleurodicus dispersus*) on three cassava (*Manihot esulanta* crantz) genotypes. Journal of Crop Research. 2003; 26(1):145-149.

8. Banjo AD, Hassan AT, Ekanayake IJ, Dixon AGO, Jackal LEN. Effect of *Aleurodicus dispersus* Russel (Spiralling whitefly) on growth indices and yield of three genotypes of cassava (*Manihot esculanta* crantz). Journal of Crop Research. 2004; 5(2-3):252-260.

9. Boopathi T. Monitoring and management of pest complex of fruits, vegetables and spices in Mizoram. In: Ann Rep of ICAR Research Complex for NEH Region 2008-2009, ICAR Research Complex for NEH Region, Umiam, Meghalaya, 2008, 289.

10. Boopathi T. Biological Control and Molecular Characterization of Spiralling Whitefly, Aleurodicus dispersus Russell on Cassava and Brinjal. Ph.D. Dissertation submitted to Tamil Nadu Agricultural University, Coimbatore, Tamil Nadu, India, 2013, 236.

11. Boopathi T, Karuppuchamy P. Evaluation of ecofriendly agents against spiralling whitefly, *Aleurodicus dispersus* Russell on brinjal. Madras Agricultural Journal. 2013; 100(4-6):559-661.

12. Boopathi T, Karuppuchamy P, Kalyanasundaram M, Mohankumar S, Ravi M. Relative efficacy of botanicals, an organic salt and fish oil rosin soap in controlling the spiraling whitefly, *Aleurodicus dispersus* Russell on cassava. Hexapoda. 2012; 19(2):59-66.

13. Boopathi T, Karuppuchamy P, Kalyanasundaram M, Mohankumar S,

Ravi M. Pathogenicity, ovicidal action, and median lethal concentrations (LC_{50}) of entomopathogenic fungi against exotic spiralling whitefly, Aleurodicus dispersus Russell. Journal of Pathogens. 2013; 7. http://dx. doi.org/10.1155/2013/393787.

14. Boopathi T, Mohankumar S, Karuppuchamy P, Kalyanasundaram M, Ravi M, Preetha B *et al.* Genetic evidence for diversity of spiralling whitefly, *Aleurodicus dispersus* (Hemiptera: Aleyrodidae) populations in India. Florida Entomologist. 2014a; 97(3):1115-1122.

15. Boopathi T, Karuppuchamy P, Kalyanasundaram M, Mohankumar S, Ravi M, Singh SB. Effect of botanicals, fish oil rosin soap and organic salt on eggs of spiralling whitefly, Aleurodicus dispersus. Indian Journal of Plant Protection. 2014b; 42(1):86-88.

16. Boopathi T, Karuppuchamy P, Kalyanasundaram M, Mohankumar S, Ravi M, Singh SB. Toxicity of newer insecticides againts spiralling whitefly, *Aleurodicus dispersus* under laboratory conditions. Indian Journal of Plant Protection. 2014; 42(2):178-180.

17. Boopathi T, Karuppuchamy P, Kalyanasundaram M, Mohankumar S, Ravi M, Singh SB. Microbial control of the exotic spiralling whitefly, *Aleurodicus dispersus* (Hemiptera: Aleyrodidae) on eggplant using entomopathogenic fungi. African Journal of Microbiological Research. 2015a; 9(1):39-46.

18. Boopathi T, Karuppuchamy P, Singh SB, Kalyanasundaram M, Mohankumar S, Ravi M. Microbial control of the invasive spiraling whitefly on cassava with entomopathogenic fungi. Brazilian Journal of Microbiology. 2015b; 46(4):1077-1085.

19. Byrne DN, Bellows TS, Parella MP. Whiteflies in agricultural system In Whiteflies-their bionomics, pest status and management. Gerling, D. (ed). Wimborne U.K. Intercept, 1990, 227-261.

20. Charati SN, Pokharkar DS, Ghorpade SA. Abundance of spiralling whitefly, a newly introduced pest in Maharashtra State. Journal of Maharashtra Agricultural University. 2003; 28:83-84.

21. D'Almeida YA, Lys JA, Neuenschwander P, Ajuonu O. Impact of Two Accidentally Introduced *Encarsia* Species (Hymenoptera: Aphelinidae) and Other Biotic and Abiotic Factors on the Spiralling Whitefly *Aleurodicus dispersus* (Russell) (Homoptera: Aleyrodidae), in Benin, West Africa. Biocontrol Science and Technology. 1998; 8(1):163-173.

22. David BVK. Regu. *Aleurodicus dispersus* Russell (Aleyrodidae: Homoptera), a whitefly pest new to India, Pestology. 1995; 19(3):5-7.

23. Esguerra NM. The spiralling whitefly Aleurodicus dispersus Russell. Entomology Bulletin. 1987; 1:1.

24. Gajendra Babu B, David PMM. New host plant records and host range of the spiralling whitefly, *Aleurodicus dispersus* Russell. (Hemiptera: Aleyrodidae). Madras agricultural Journal. 1999; 86:305-313.

25. Geetha B. Biology and Management of Spiralling Whitefly *Aleurodicus dispersus* Russell (Homoptera: Aleurodidae). *Ph.D Dissertation*, Tamil Nadu Agricultural University, Coimbatore, India, 2000, 196.

26. Geetha B, Loganathan M, Swamiappan M. Record of spiralling whitefly, *Aleurodicus dispersus* Russell on groundnut. Insect Environment. 1998; 4(2):55.

27. Geetha B, Swamiappan M, Loganathan M. New hosts for spiralling whitefly, *Aleurodicus dispersus* Russell in Tamil Nadu. Insect Environment. 1999; 5(20):80.

28. Gerling D. Natural enemies of *Bemisia tabaci*, biological characteristics and potential as biocontrol agents: a review, Agriculture, Ecosystems and Environment. 1986; 17:99-110.

29. Hernandez-Suarez E, Carnero A, Aguilar A, Prinsloo G, Laslle J, Polaszek A. Parasitoids of whiteflies (Hymenoptera: Aphelinidae, Eulophidae, Platygastridae; Hemiptera: Aleyrodidae) from the Macaronesian archipelagos of the Canary Islands, Madeira and the Azores. Systematics and Biodiversity. 2003; 1:55-108.

30. Kajita H, Alam MZ. Whiteflies on guava and vegetables in Bangladesh and their Aphelinid parasitoids. Applied Entomology Zoology. 1996; 31:159-162.

31. Kajita H, Samudra IM, Naito A. Discovery of the spiralling whitefly *Aleurodicus dispersus* Russell (Homoptera: Aleyrodidae) from Indonesia, with notes on its host plants and natural enemies, Applied Entomology and Zoology. 1991; 26:397-400.

32. Kavitha Kirubavathy, Saranya and Mariamma Ninan. Neem for the management of whiteflies on tapioca. In National Symposium on Biological Control of insects in agricultural, forestry, medicine and veterinary sciences. Department of Zoology, Bharathiar University, Coimbatore, Tamil Nadu, India. 1999; 98:21-22.

33. Kumar K, Singh R, Lal HJ. Control of spiralling whitefly. In Report presented on, Annual Research Meeting of Ministry of Primary Industries, Fiji, 1987.

34. Kumashiro BR, Lai PY, Funasaki GY, Teramoto KK. Efficacy of *Nephaspis amnicola* and *Encarsia haitiensis* in controlling Aleurodicus dispersus in Hawaii, Proceedings Hawaiian entomological Society. 1983; 24(2&3):261-269.

35. Laprade S, Gerdas VH. Management of the spiraling whitefly *Aleurodicus dispersus* Russell (Homoptera, Aleyrodidae) in banana (*Musa* AAA) using insecticides and oil. Corbanana. 1998; 22(48):89-94.

36. Mallappanavar MC. Bioecology and management of spiraling whitefly *Aleurodicus dispersus* Russell by *Verticillium lecanii* (Zimm.) on guava. M.Sc. (Agri.) thesis, University of Agricultural Sciences, Dharwad, India, 2000.

37. Mani M, Krishnamoorthy A. Spiralling whitefly in horticultural crops. The Hindu. 1997a; 20(19):28.

38. Mani M, Krishnamoorthy A. Discovery of Australian ladybird beetle (*Cryptolaemus montrouzieri*) on spiralling whitefly (Aleurodicus dispersus) in India. Insect Environment. 1997b; 3(1):5-6.

39. Mani M, Krishnamoorthy A. Natural enemies and host plants of spiralling whitefly *Aleurodicus dispersus* Russell (Homoptera: Aleyrodidae) in Bangalore, Karnataka, Entomon. 1999a; 24(1):75-80.

40. Mani M, Krishnamoorthy A. *Chilocorus nigrita* on spiralling whitefly. Insect Environment.1999b; 4(4):118-119.

41. Mani M, Krishnamoorthy A. Predatory potential and development of Australian ladybird beetle, *Cryptolaemus montrouzieri* Muls. On the spiralling whitefly, *Aleurodicus dispersus* Russell. Entomon. 1999a; 24:166-171.

42. Mani M, Krishnamoorthy A. Population dynamics of spiralling whitefly, *Aleurodicus dispersus* Russell (Aleyrodidae: Homoptera) and its natural enemies on guava in India. Entomon, 2000; 25:29-34.

43. Mani M, Krishnamoorthy A. Classical biological control of spiralling whitefly, *Aleurodicus dispersus* Russell-An appraisal. Insect Science and its Application. 2002; 22:263-273.

44. Mani M, Dinesh MS, Krishnamoorthy A. Biological control studies on the spiralling whitefly, *Aleurodicus dispersus* Russell. (Aleyrodidae:

Homoptera). In Entomology Congress, Trivandrum, Kerala, India, 2000.

45. Mani M, Ragunatha R, Dinesh MS, Krishnamoorthy A. Spiralling whitefly *Aleurodicus dispersus* Russell in Hyderabad. Insect Environment. 2001; 7(2):82.

46. M'Boob SS, van Oers CCCM. Spiralling whitefly (*Aleurodicus dispersus*): a new problem in Africa. FAO Plant Protection Bulletin. 1994; 42:59-62.

47. Megir-Gumbek. Study on control of white fly. Annual Report of the Research Branch, Department of Agriculture for the year Sarawak: Department of Agriculture, 1986, 1987, 83-84.

48. Mikunthan G. Utilization of fungal pathogens for the biological control of thrips (*Scirtothrips dorsalis* Hood) and mite {*Polyphagotarsonemus latus* (Banks)} in chilli (*Capsicum annum* L.). *Ph. D Thesis*, University of Agricultural Sciences, Bangalore, 2004, 203.

49. Mound LA, Halsey SH. Whitefly of the World. A systematic catalogue of the Aleyrodidae (Homoptera) with Host Plant and Natural Enemy Data. British Museum (Natural History). John Wiley & Sons, London, 1978, 340.

50. Muniappan R. Spiraling whitefly threat, *The Hindu*, dated, 1996, 12.

51. Muralikrishna M. Bioecology, host range and management of spiraling whitefly, *Aleurodicus disperses* Russell (Homoptera: Aleyrodidae). M.Sc. (Agri.) thesis, University of Agricultural Sciences, Bangalore, 1999, 67.

52. Mware B, Narla R, Amata R, Olubayo F, Songa J, Kyamanyua S *et al.* Efficiency of cassava brown streak virus transmission by two whitefly species in coastal Kenya. Journal of General and Molecular Virology. 2009; 1(4):040-045.

53. Natarajan K, Sundaramurthy VT, Chidambaram P. Usefulness of fish oil resin soap in the management of whitefly and other sap feeding insects on cotton. Entomon. 1991; 16(3):229-232.

54. Nechols JR. Entomology: biological control. *Annual Report.* Guam Agricultural Experimental Station, 1982, 33-49.

55. Nechols JR. Entomology: biological control. *Annual Report* 1983. Guam Agricultural Experimental Station, 1983, 26-27.

56. Nijhof BW, Oudman L, Torres R, Garrido C. The introduction of

Encarsia guadeloupae (Hymenoptera: Aphelinidae) for control of *Aleurodicus dispersus* and *Lecanoideus floccissimus* (Homoptera: Aleurodidae) on Tenerife. In Proceedings of the Section Experimental and Applied Entomology of the Netherlands Entomological Society. 2000; 11:41-47.

57. Palaniswami MS, Pillai KS, Nair RR, Mohandas C. A new cassava pest in India, Cassava Newsletter. 1995; 19:6-7.

58. Paulson GS, Kumashiro BR. Hawaiian Aleyrodidae, In Proceedings of Hawaiian Entomological Society. 1985; 25:103-129.

59. PDBC. Annual Report of 1999-2000. Project Directorate of Biological Control, Banglaore, Karnataka, 2000, 216.

60. Ragumoorthy KN, Kempraj T. Sucking pests of cassava, The Hindu. 1996; 119(123):28.

61. Ramani S. Fortuitous introduction of an aphelinid parasitoid of the spiralling whitefly, *Aleurodicus dispersus* Russell (Homoptera: Aleyrodidae), in to the Lakshadweep Islands with notes on host plants and other natural enemies. Journal of Biological Control. 2000; 14:55-60.

62. Ranjith AM. Mealy whitefly, pest with a wide host range, *The Hindu*, dated. 1998; 29:4.

63. Ranjith AM, Rao DS, Thomas J. New host records of the Mealy Whitefly, *Aleurodicus dispersus* Russell in Kerala. Insect Environment. 1996; 2(2):35-38.

64. Ranjit AM, Thomas J, Ramesha B, Sreekala NV, Sajithakumari Sitaramarao D. Management strategies for the mealy whitefly, *Aleurodicus dispersus* Russell (Homoptera: Aleyrodidae). In: First National Symposium on Pest Management in Horticultural Crops: Environmental Implications and Thrusts. Association for Advancement of Pest Management in Horticultural Ecosystems, Bangalore, Karnataka, 1997.

65. Russell LM. A new species of *Aleurodicus* Douglas and two close relatives, The Florida Entomologist. 1965; 48:47-55.

66. Srinivasa MV. Host plants of the spiralling whitefly, *Aleurodicus dispersus* (Hemiptera: Aleyrodidae), Pest management in Horticultural Ecosystem. 2000; 6:79-105.

67. Srinivasa MV, Viraktamath CA, Reddy C. A new parasitoid of the

spiralling whitefly *Aleurodicus dispersus* Russell (Hemiptera: Aleyrodidae) in South India, Pest Management in Horticultural Ecosystems. 1999; 5(1):59-61.

68. Venugopal Rao N, Reddy AS, Reddy PS. Relative efficacy of some new insecticides on insect pests of cotton. Indian Journal of Plant Protection. 1990; 18:53-58.

69. Waterhouse DF, Norris KR. Aleurodicus dispersus Russell. In Biological Control: Pacific Prospects-Supplement 1, ACIAR Monograph No.12, ACIAR, Canberra, Australia, 1989, 13-22.

70. Wen HC, Hsu TC, Chen CN. Yield loss and control of spiralling whitefly (*Aleurodicus dispersus* Russell). Journal of Agricultural Research, China. 1995; 41:147-156.

71. Wijesekera GAW, Kudagamage C. Life history and control of 'spiralling' white fly *Aleurodicus dispersus* (Homoptera: Aleyrodidae): fast spreading pest in Sri Lanka. Quarterly Newsletter-Asia and Pacific Plant Protection Commission. 1990; 33(2):22-24.

72. Wraight SP, Carruthers RI, Bradley CA, Jaronski ST, Lacey LA. Wood P. n Galaini-Wraight S. Pathogenicity of the entomopathogenic fungi *Paecilomyces* spp. and *Beauveria bassiana* against the silverleaf whitefly, Bemisia argentifolii. Journal of Invertebrate Pathology. 1998; 71:217-226.

73. Wraight SP, Carruthers RI, Jaronski ST, Bradley CA, Garza CJ, Galaini-Wraight S. Evaluation of the entomopathogenic fungi *Beauveria bassiana* and *Paecilomyces fumosoroseus* for microbial control of the silverleaf whitefly, Bemisia argentifolii. Biological Control. 2000; 17:203-217.

Chapter - 4
Microbial biopesticides Based Insect Pest Management in Cotton-Present Status and Futuristic Approach

<u>Authors</u>

Arunkumar N
Department of Agricultural Microbiology, Tamil Nadu Agricultural University, Coimbatore, Tamil Nadu, India

Selvi C
Department of Agricultural Entomology, Tamil Nadu Agricultural University, Coimbatore, Tamil Nadu, India

Guru Prasanna Pandi G
Division of Crop Protection, ICAR-National Rice Research Institute, Cuttack, Odisha, India

Raghu S
Division of Crop Protection, ICAR-National Rice Research Institute, Cuttack, Odisha, India

Sivakumar G
Division of Insect Ecology, ICAR-NBAIR, Bangalore, Karnataka, India

Chapter - 4

Microbial Biopesticides Based Insect Pest Management in Cotton-Present Status and Futuristic Approach

Arunkumar N, Selvi C, Guru Prasanna Pandi G, Raghu S and Sivakumar G

1. Introduction

Microbial pest control is concerned with the employment of microorganisms for the control and reduction of harmful animals, plants and insects. Pesticides based on microorganisms and their products have proven to be highly effective, species specific and eco-friendly in nature, leading to their adoption in pest management strategies around the world. The microbial biopesticide constitutes about 90% of total biopesticides market and there is ample scope for further development in agriculture and public health. This review is aimed at highlighting the roles of microbes (bacteria, fungi, viruses and some microscopic protozoan) in biological pest control of insect pests of cotton, stating their nature and their application benefits. Cotton pest management is mainly affected due to insecticide resistance and tolerance developed by the pests, which led to excessive use of insecticides on the crop. Considering the serious impacts of chemical pesticides on our environment and the general public health, it has become necessary to emphasize on the importance of microbial pest so as to reduce the reliance on chemical pesticides. Microbial pesticides such as *Bacillus thuringiensis*, wax degrading bacteria, entomopathogenic fungi, entomopathogenic protozoans and nematodes helps to control insect pests in eco friendly way, reducing environmental and public safety hazards of chemicals.

For the past two decades, there has been an increased interest in biological control agents for insect pest control. More number of biocontrol agents was screened for their efficacy and environmental impact including mammalian safety. Many organisms have been investigated as potential agents for pest control, including viruses, fungi, bacteria, protozoa, nematodes, invertebrate predators and fish. Pesticides that include insecticides, herbicides, and fungicides are employed in modern agriculture to control pests and to increase crop yield. In both developed and developing countries, the use of chemical pesticides has increased dramatically during the last few decades. Control of pests with synthetic chemicals results in several problems. The residues of these synthetic insecticides cause toxic effects on wild life (e.g., birds, beneficial insects like honeybees). These chemical insecticides also induce harmful chemical changes on non-target insects/pests on their predators, parasites, etc. They can also be harmful to humans and domestic animals. Other environmental concern is the contamination of ground water (Lacey, 2000). Despite many years of effective control by conventional agrochemical insecticides, a number of factors are threatening the effectiveness and continued use of these agents. These include the development of insecticide resistance and use-cancellation or de-registration of some insecticides due to human health and environmental concerns.

Particularly in cotton, past two decades, the perplexities in pest management intensified with more and more insect species developing resistance to insecticides. Cotton pest management was particularly affected due to insecticide resistance, which was a consequence of excessive use of insecticides on the crop. Insecticide resistance rendered insecticides ineffective, thus increasing the need for repeated applications, wastage of resources and consequent environmental pollution (Kranthi, 2007). Therefore, an eco-friendly alternative is the need of the hour. Improvement in pest control strategies represents one method to generate higher quality and greater quantity of agricultural products. Therefore, there is a need to develop biopesticides which are effective, biodegradable and do not leave any harmful effect on environment (Nicholson, 2007). Microbial control agents, based on naturally occurring fungi, bacteria, viruses or nematodes have offered some realistic alternatives to chemical pesticides when used as part of an ecologically based integrated pest management (EBIPM) or area-wide pest management strategy (AWPM). There are many reasons for the recent increased interest in microbial biopesticides, including the development of resistance to conventional synthetic pesticides, a decline in the rate of discovery of novel insecticides, increased public perception of the dangers associated with synthetic pesticides, host-specificity of microbial pesticides and improvement in the production and formulation technology of microbial biopesticides. In view of various opportunities and challenges that are associated with the development of microbial biopesticides, this article reviews the most important microbial biopesticides and their effectiveness against sucking pests of cotton (Koul, 2011).

2. Insect Pests of Cotton

Cotton (*Gossypium hirsutum* L.) is one of the most important commercial fibre crops of global significance and a major source of raw material for the domestic textile industry in India. India ranks first in terms of cultivated area, occupying 32% of the global cotton area followed by China and contributes 21% of the global cotton produce (36.10 mln bales), ranking second to China. The productivity of cotton in India is about 537 kg ha^{-1}, whereas, Brazil holds highest productivity level of 2,027 kg ha^{-1}among the major cotton growing countries. The major cotton growing states in India are Maharashtra (7.36 M ha), which account for 37% of the area followed by Gujarat, Andhra Pradesh, Punjab, Haryana, Rajasthan, Madhya Pradesh, Karnataka and Tamil Nadu (0.57 M ha).The cotton crop can be ravaged by several insect pests and nematodes causing drastic reductions in yield. The pest spectrum of cotton is quite complex. About 1,326 insect pests in all over the world. In India about 162 pest was recorded in cotton.

Cotton is infested by a large number of insect pests right from the sowing till harvest. Sap sucking insects like *Empoasca devastans* (Ishida), *Aphis gossypii* (Glover), *Thrips tabaci (*Lind), *Bemisia tabaci* (Gennadius), Cotton mealybug *Phenacoccus solenopsis* (Tinsley), Papaya mealybug *Paracoccus marginatus* (Williams and Granara de Willink), Spherical mealybug *Nipaecoccus viridis* (Newstead), Pink hibiscus mealybug *Maconellicoccus hirsutus* (Green), Striped mealybug *Ferrisia virgata (Cockerell)*Malvastrum mealybug *Ferrisia malvastra* (McDaniel), Mango mealybug *Rastrococcus iceryoide* (Green), Mealybug *Drosichiella* spp. Indian cotton mirid *Creontiades biseratense* (Distant), Mirid *Campylomma livida* (Reuter), Mirid *Hyalopeplus lineifer* (Walker), T Mosquito bug *Helopeltis bradyi* (Waterhouse) and cotton boll worms including American bollworm *Helicoverpa armigera* (Hubner), Spotted bollworms *Earias insulana* (Boisduvel), *E. vittella* (Fab.), Pink bollworm *Pectinophora gossypiella* (Saunders) cause serious problem and resulting reduction in yield and quality of cotton (Nagrare *et al*., 2013).

3. Microbial Pesticides

Microbial pesticides are biological pesticides, developed from naturally occurring or genetically altered bacteria, fungi, algae, viruses or protozoans, either the organisms as such or a toxin derived from the microorganisms. A microbial toxin can be defined as a biological toxin material derived from a microorganism, such as a bacterium or fungus. Pathogenic effect of those microorganisms on the target pests are so species specific. The effect by microbial entomopathogens occurs by invasion through the integument or gut of the insect, followed by multiplication of the pathogen resulting in the death of the host, e.g., insects. Studies have demonstrated that the pathogens produce insecticidal toxin important in pathogenesis. Most of the toxins produced by microbial pathogens which have been identified are peptides, but they vary greatly in terms of structure, toxicity and specificity (Burges, 1981). Microorganism bacterium, fungus, virus or protozoan as the active ingredient can control many different kinds of pests, although each separate active ingredient is relatively specific for its target pest. For example, there are fungi that control certain weeds, and other fungi that kill specific insects. One bacterial species like *Bacillus thuringiensis* may be more effective on *Aedes aegypti* while one another *B. sphaericus* strain can be effective on different types of insects and pests (Lacey, 2001) but in general they are target specific and ecofriendly.

4. Biological Pesticides

The biopesticides are certain types of pesticides derived from natural materials, such as animals, plants, bacteria, and certain minerals. Biopesticides are biochemical pesticides that are naturally occurring substances that control pests by nontoxic mechanisms. They are living organisms (natural enemies) or their products (phyto- chemicals, microbial products) or byproducts (semiochemicals) which can be used for the management of pests that are injurious to crop plants. Biopesticides have an important role in crop protection, although most commonly in combination with other tools including chemical pesticides as part of Bio-intensive Integrated Pest Management. Biological pesticides based on pathogenic microorganisms specific to a target pest offer an ecologically sound and effective solution to pest problems. They pose less threat to the environment and to human health. The most commonly used biopesticides are living organisms, which are pathogenic for the pest of interest. These include biofungicides (*Trichoderma*), bioherbicides (*Phytopthora*) and bioinsecticides (*Bacillus thuringiensis*, *B. sphaericus*) (Gupta and A.K. Dikshit, 2010).

5. Advantages of Microbial Insecticides for Insect Pest Control in Cotton

Chemical control has not provided a long-term solution for cotton pest problems because of the high costs, environmental impact and related problems (insecticide-resistant insect strains, reduction of natural enemies of pests, the resurgence of pest populations in the absence of natural enemies and the occurrence of secondary pests). Individual products differ in important ways, but the following list of beneficial characteristics applies to microbial insecticides in general. The organisms used in microbial insecticides are essentially nontoxic and nonpathogenic to wildlife, humans, and other organisms not closely related to the target pest. The safety offered by microbial insecticides is their greatest strength, the toxic action of microbial insecticides is often specific to a single group or species of insects, and this specificity means that most microbial insecticides do not directly affect beneficial insects (including predators or parasites of pests) in treated areas. If necessary, most microbial insecticides can be used in conjunction with synthetic chemical insecticides because in most cases the microbial product is not deactivated or damaged by residues of conventional insecticides. (Follow label directions concerning any limitations.) Because their residues present no hazards to humans or other animals, microbial insecticides can be applied even when a crop is almost ready for harvest. In some cases, the pathogenic microorganisms can become established in a pest population or its habitat and provide control during subsequent pest generations or seasons. They also enhance the root and plant growth by way of encouraging the beneficial soil microflora. By this way microbial insecticides take a part in the increase of the crop yield.

6. Microbial Biopesticides against Insect Pests of Cotton

Out of all the biopesticides used today, microbial biopesticides constitute the largest group of broad spectrum biopesticides, which are pest specific (i.e. do not target non pest species and are environmentally benign). There are at least 1500 naturally occurring insect-specific microorganisms, 100 of which are insecticidal (Khachatourians, 2009). Over 200 microbial biopesticides are available in 30 countries affiliated to the Organization for Economic Co-operation and Development (OECD). There are 53 microbial biopesticides registered in the USA, 22 in Canada and 21 in the European Union (EU) although reports of the products registered for use in Asia are variable (Thakore, 2006). Overall, microbial biopesticide registrations are increasing globally, the expansion of various technologies has increased the scope for more products and the change in the trend to develop microbial products is definitely on the rise. Microbial biopesticides that are applied for the control of sucking pests of cotton are of different types such as entomopathogenic fungi, viral pesticides, bacterial biopesticides including Bt and recently wax degrading bacteria.

6.1 Entomopathogenic Fungi

Entomopathogenic fungi are important natural regulators of insect populations and have potential as mycoinsecticide agents against diverse insect pests in agriculture. These fungi infect their hosts by penetrating through the cuticle, gaining access to the hemolymph, producing toxins, and grow by utilizing nutrients present in the haemocoel to avoid insect immune responses (Meadows, 1993). Entomopathogenic fungi may be applied in the form of conidia or mycelium which sporulates after application. The use of fungal entomopathogens as alternative to insecticide or combined application of insecticide with fungal entomopathogens could be very useful for insecticide resistant management (Hajeck and Leger, 1994). Entomopathogenic fungi belong to 12 classes within six phyla and belong to four major groups; Laboulbeniales, Pyrenomycetes, Hyphomycetes and Zygomycetes. Some of the most widely used species include *Beauveria bassiana, Metarhizium anisopilae, Nomuraea rileyi, Paecilomyces farinosus* and *Verticillium lecanii.* The commercial mycoinsecticide 'Boverin' based on *Beauveria bassiana (Bals.)* with reduced doses of trichlorophon have been used to suppress the second-generation outbreaks of *Cydia pomonella.* Anderson *et al.* (1989) detected higher insect mortality when *B. bassiana* and sublethal concentrations of insecticides were applied to control Colorado potato beetle (*Leptinotarsa decemlineata*), attributing higher rates of synergism between two agents (Ferron, 1971). Toxins of *B. bassiana* such as Beauvericin, Oosporin weaken the host's immune system. Its victims are mainly caterpillars, plant hoppers and also beetles, whiteflies, aphids, thrips, locusts, termites and mites (Kreutz *et al.,* 2004). The fungus *V. lecanii* infects mostly whiteflies, thrips and rarely caterpillars (Jung *et al.,* 2006). Even very low concentration of *Beauveria bassiana Bals.* And *Verticillium lecanii Zimm* (0.25 percent) recorded the highest mortality of *H. armigera* (86.67 %), *S. litura* (86.67 %), *E. vittella* (73.33 %), *A. gossypii* (100.00 %), *B. tabaci* (100.00 %) and *A. devastans* (93.33 %) respectively, under laboratory conditions (Karthikeyan and Selvanarayanan, 2011).

6.2 Viral Pesticides

There are more than 1600 different viruses which infect 1100 species of insects and mites. A special group of viruses, called baculovirus, to which about 100 insect species are susceptible, accounts for more than 10 percent of all insect pathogenic viruses. Baculoviruses are rod- shaped particles which contain DNA. Most viruses are enclosed in a protein coat to make up a virus inclusion body. Alkaline condition of insect's midgut dissolves the protein covering and the viral particles are released from the inclusion body. These particles fuse with the midgut epithelial cells, multiply rapidly and eventually kill the host. But, viral pesticides are more expensive than chemical agents. Furthermore, many baculoviruses are host specific. Therefore they cannot be used to control several different pests. The action of baculoviruses on insect larvae is too slow to satisfy farmers. These viral preparations are not stable under the ultraviolet rays of the sun. Efforts are being made to encapsulate baculoviruses with UV protectants to ensure a longer field life (Usta, 2013).

NPVs and GVs are used as pesticides but the group based on nucleopolyhedrosis viruses is much larger. The first viral insecticide Elcar™ was introduced by Sandoz Inc. in 1975Elcar™ was a preparation of *Heliothis zea* NPV which is relatively broad range baculovirus and infects many species belonging to genera *Helicoverpa* and *Heliothis*. HzSNPV provided control of not only cotton bollworm, but also of pests belonging to these genera attacking soybean, sorghum, maize, tomato and beans. In 1982 Sandoz decided to discontinue the production. The resistance to many chemical insecticides including pyrethroids revived the interest in HzSNPV and the same virus was registered under the name GemStar™. HzSNPV is a product of choice for biocontrol of *Helicoverpa armigera*]. Countries with large areas of such crops like cotton, pigeonpea, tomato, pepper and maize, e.g. India and China, introduced special programs for the reduction of pests by applying Viral pesticides (James, 1984). Commercial preparations such as SPOD-X™ containing Spodoptera exigua NPV are very effective in control of insects on vegetable crops and Spodopterin™ containing Spodoptera littoralis NPV are used to protect cotton, corn, and tomatoes (Szewczyk et al. 2006). In China, twelve baculoviruses have been authorized as commercial insecticides (Sun and Peng, 2007), including *H. armigera* NPV (the most widely used virus in China for cotton, pepper, and tobacco protection), *Spodoptera litura* NPV (vegetables), *S. exigua* NPV (vegetables), *Buzura suppressaria* NPV (tea), *Pieris rapae* GV, and *Plutella xylostella* GV (vegetables).

6.3 Bacteria as Insect Control in Cotton

The year 2011 was marked as a century of research and development of *Bacillus thuringiensis* (Bt). Bt is a Gram-positive, rod-shaped, spore-forming entomopathogenic bacterium characterized at the species level by the production upon sporulation of a parasporal inclusion body, the crystal which is toxic to insects and other invertebrates and is the most commonly used commercial biopesticide worldwide (Bravo *et al.*, 2011). Bt protein toxins are highly selective to their target insect, are innocuous to humans, vertebrates, plants and are completely biodegradable. Therefore, Bt is a viable alternative for the control of insect pests in agriculture and disease spreading vectors in public health. The primary sources of proteins were used in developing genetically engineered crops to protect them from insect damage. Transgenic crops based on insecticidal crystal proteins of Bt are now an international industry with revenues of several billion dollars per year (James, 2011).

Classification of Bt strains has been by H serotyping, the immunological reaction to the bacterial flagella antigen. Specific flagellin amino acid sequences have been correlated to specific Bt H serotypes and at least 69 H serotypes and 82 serological varieties (serovars) of Bt have been characterized from around the world. *Bacillus thuringiensis* (Bt) based biopesticides are especially of utmost importance and occupy almost 97 per cent of the world biopesticides market. But the substantial use of bacterial biopesticides is limited because a biopesticide can be effective only if it has a potential major impact on the target pest, substantial market size, variability of field performance and cost effective (Brar *et al.*, 2006). The lepidopteran are the most important pest of cotton, and the efficacy of Bt based products is similar to that of chemicals (14% for chemicals vs. 15.13 % for biopesticides) (Jiménez, 1996). But the application of Bt based biopesticides is gaining importance in control of severe disease causing mosquitoes than their application in the control of crop pests.

6.4 Wax Degrading Microorganisms against Wax Coated Sucking Pests of Cotton

Mealybugs are a major group of waxy coated insects which fall under the order Hemiptera, sub order Sternorrhyncha. The suborder Sternorrhyncha contains four superfamilies: Psylloidea (jumping plant lice); Aleyrodoidea (whiteflies); Aphidoidea (aphids) and Coccoidea (scale insects and mealybugs). Mealybugs are very soft-bodied, sap feeders with mouthparts adapted to piercing and sucking and secrete powdery covers all over the body. Wax deposits were maximum and thickest on the adult females, particularly the oldest. The powdery waxy coating protects the mealybugs from desiccation and penetration of toxic chemicals and pathogens through the integument (Watson and Kubiriba, 2005). But there are certain hydrolyzing microbes termed as the wax degrading microbes, are capable of degrading this waxy coating on mealybugs and certain other insects. Wax degrading microorganisms are a distinct group of microorganisms capable of degrading waxy substances through pseudosolubilization, adherence, biosurfactant production, enzyme secretion (lipase) and adsorption of waxes by direct contact (Arunkumar *et al.*, 2016).

Certain wax degrading bacteria (WDB) produce these heterogeneous surface active amphiphilic molecules or biosurfactants to degrade compex hydro carbons like waxes. Wax degrading microbes also produce cell wall degrading enzymes such as lipases as a lethal component against insect pests. The lipase enzyme plays a major role in the separation and hydrolysis of the insect waxy mass making the cuticle accessible to degradation by chitinases and proteases (Arunkumar *et al.*, 2017a). The mortality of *Ceratovacuna lanigera* (wolly aphid) of sugarcane under laboratory conditions increased several fold due to the microbial lipases of *Myrothecium verrucaria*, *Metarhizium anisopliae* and of *Bacillus* sp. B1. *Myrothecium verrucaria* produces chitinase, lipase, protease and β-1, 3-glucanase (Shaikh and Deshpande, 1993) that killed both first (I) and fourth (IV) instar *Aedges aegypti* mosquito larvae, within 48 h of application (Mendonsa *et al.* 1996). Entomopathogenous fungi such as *Beauveria bassiana* and *Metarhizium anisopliae* that can grow on long chain alkanes (up to C33) as well as many fatty acids, alcohols, and triacylglycerides were reported in many literatures.

An excellent report on the efficacy of wax degrading bacteria on insect pests control was by Salunkhe *et al.* (2013), in their investigation against polyphagous insect pest, the pink mealybug *Maconellicoccus hirsutus*. The bacterial isolates originating from *M. hirsutus* cadavers were screened for their wax degrading effects and the three most potent wax degrading isolates were identified to be *Serratia marcescens*, *Pseudomonas aeruginosa* and *Bacillus subtilis*. Application of these wax degrading bacterial isolates reduced female longevity, offspring production as well as reduction in weight and wax content of emerging mealybug adults. Eight strains of wax degrading bacteria identified as *Pseudoxanthomonas* sp. PSAD1, *Acinetobacter* sp. PSAD2 and PSAD 7, *Klebsiella* sp. PSAD3 and PSAD8, *Providencia* sp. PSAD5 and *Serratia* Sp. PSAD9 capable of producing biosurfactants and lipase enzyme was isolated from the carcass of mealybug species *Phenacoccus solenopsis* (Tinsley) and *Ferrisia virgata* (Cockerell) (Arunkumar *et al.*, 2017 b). At present there is no commercial formulation of wax degrading bacteria for the control of wax coated sucking insect pests of cotton but research is underway by the authors of this review to develop a novel biopesticides formulation of wax degrading bacteria.

6.5 Protozoa and Entamopathogenic Nematodes

Protozoan pathogens naturally infect a wide range of insect hosts. Although these pathogens can kill their insect hosts, many are more important for their chronic, debilitating effects. One important and common consequence of protozoan infection is a reduction in the number of offspring produced by infected insects. Although protozoan pathogens play a significant role in the natural limitation of insect populations, few appear to be suited for development as insecticides (Hoffmann and Frodsham, 1993).

Nematodes are microscopic multicellular round worms, Colorless, unsegmented, and lacking appendages, nematodes may be free-living, predaceous, or parasitic. Many of the parasitic species cause important diseases of plants, animals, and humans. Other species are beneficial in attacking insect pests, mostly sterilizing or otherwise debilitating their hosts. Nematodes used for insect control infect only insects or related arthropods; they are called entomogenous nematodes (Moscardi, 1999). The only insect-parasitic nematodes possessing an optimal balance of biological control attributes are entomopathogenic or insecticidal nematodes in the genera *Steinernema* and *Heterorhabditis*.

The entomogenous nematodes *Steinernema feltiae* (sometimes identified as *Neoaplectana carpocapsae*), *S. scapteriscae*, *S. riobravis*, *S. carpocapsae* and *Heterorhabditis heliothidis* are the species most commonly used in insecticidal preparations. Within each of these species, different strains exhibit differences in their abilities to infect and kill specific insects. In general, however, these nematodes infect a wide range of insects. On a worldwide basis, laboratory or field applications have been effective against over 400 pest species, including numerous beetles, fly larvae, and caterpillars. A number of cotton pests are found susceptible to entomopathogenic nematodes including bollworm, *H.zea*, fall armyworm, *Spodoptera frugiperda* (J.E.Smith), beet armyworm, *S. exigua* (HUbner), cabbage looper, *Trichoplusia ni* (Hubner), tobacco budworm, *Heliothis virescens* (Fabricius) and the pink bollworm, *Pectinophora gossypiella* Saunders (Gaugler,1999). But still the application of these entomopathogenic nemeatodes is underscore because a very few cause insect death but also these species tend to be difficult (e.g. tetradomatids) or expensive (e.g. mermithids) to mass produce, have narrow host specificity against pests of minor economic importance, possess modest virulence (e.g., sphaeruliids) or are otherwise poorly suited to exploit for pest control purposes.

Conclusion

Extensive use of chemical pest control agents are regarded as ecologically unacceptable in all countries. Now, it is time to replace them gradually with biopesticides with various biological activities that occur naturally from living organisms. Biopesticides based on microbes as whole, microbial toxins, biochemicals derived from micro-organisms are emerging faster but the potential of certain group of microbes still remains unexplored. The demand for bio-pesticides is rising steadily in all parts of the world. When used in Integrated Pest Management systems, biopesticides efficacy can be equal to or better than conventional products, especially for crops like fruits, vegetables, nuts and flowers. By combining performance and safety, biopesticides perform efficaciously while providing the flexibility of minimum application restrictions, superior residue and resistance management potential, and human and environmental safety benefits. It is very likely that in future their role will be more significant in agriculture and forestry. Biopesticides clearly have a potential role to play in development of future integrated pest management strategies. Hopefully, more rational approach will be gradually adopted towards biopesticides in the near future and short-term profits from chemical pesticides will not determine the fate of biopesticides.

References

1. Arunkumar N, Gulsar Banu J, Gopalakrishnan N, Prakash AH. Wax Degrading Bacteria: Scope and Applications in Agriculture. Int. J Curr. Microbiol. App. Sci. 2017a; 6(2):649-664.
2. Arunkumar N, Gulsar Banu J, Gopalakrishnan N, Prakash AH. Isolation, screening and characterization of microbial surfactants producing wax degrading bacteria from cotton mealybugs, *Phenacoccus solenopsis* Tinsley and *Ferrisia virgata* Cockerell (Homoptera: Pseudococcidae). Journal of Entomology and Zoology Studies. 2017b; 5(2):1191-1195.
3. Arunkumar N, Gulsar Banu J, Gopalakrishnan N, Prakash AH. Isolation and characterization of wax degrading bacteria isolated from cotton mealy bugs for their application in sustainable agriculture production. In: Fifth International Conference on Sustainable Utilization of Tropical Plant Biomass: Bioproducts, Biocatalysts and Biorefinery (Sut B4). Nov 17 & 18, TNAU, Coimbatore, Tamil Nadu, India, 2016.
4. Brar SK, Verma M, Tyagi RD. Process Biochem. 2006; 41:323-342.
5. Bravo A, Likitvivatanavong S, Gill SS. Insect Biochem Mol. 2011; 41:423-431.
6. Burges HD. Safety, Safety Testing and Quality Control of Microbial Pesticides, 1981, 176.
7. Ferron P. Modification of the development of Beauveria tenella mycosis in Melolontha melolontha larvae by means of reduced doses of Organophosphorus insecticides, Entomologia Experimentalis ET Applicata. 1981; 14:457-466.
8. Gaugler R. Matching nematode and insect to achieve optimal field

performance. In: Optimal use of insecticidal nematodes in pest management. Polavarapu S. (ed.), Rutgers University, New Brunswick, NJ, 1999, 9-4.

9. Gupta S, Dikshit AK. Biopesticides: An ecofriendly approach for pest control. Journal of Biopesticides. 2010; 3(1):186-188.

10. Hajeck AE, St. Leger Interactions between fungal pathogens and insect hosts. Annual Review of Entomology. 1994; 39:293-322.

11. Hoffmann MP, Frodsham AC. Natural Enemies of Vegetable Insect Pests. Cooperative Extension, Cornell University, Ithaca, NY, 1993, 63.

12. James C. Global Status of Commercialized Biotech GM Crops. ISAAA Brief No. 43, ISAAA, Ithaca, NY, 2009-2011.

13. James JG. Persistence of Nosema locustae Spores in Soil as Determined by Fluorescence Microscopy. Appl Environ Microbiol, February. 1984; 47(2):313-318.

14. Jiménez J. Control de Heliothis virescens (Fabr.) com medios biológicos en el cultivo Del tabaco. In: Curso International de Sanidad Vegetal en Tabaco, Cuba, INISAV, 14, 1996.

15. Jung HS, Lee HB, Kim K, Lee EY. Selection of *Lecanicillium* strains for aphid *(Myzus persicae)* control. Korean. J Mycol. 2006; 34:112-118.

16. Karthikeyan A, Selvanarayanan V. *In vitro* Efficacy of *Beauveria bassiana* (Bals.) and *Verticillium lecanii* (Zimm.) against selected insect pests of cotton. Recent Research in Science and Technology. 2011; 3(2):142-143.

17. Khachatourians GG. Insecticides and microbials. Applied Microbiology: Agro/Food, 2009, 95-109.

18. Koul O. Microbial biopesticides: opportunities and challenges. CAB Reviews: Perspectives in Agriculture, Veterinary Science, Nutrition and Natural Resources. 2011; 6:056-065.

19. Kranthi KR. Insecticide resistance management in cotton to Enhance productivity. Modern training course on "Cultivation of long staple cotton, 2007-2011.

20. Kreutz J, Zimmermann G, Vaupel O. Horizontal transmission of the entomopathogenic fungus, *Beauveria bassiana* among the spruce bark beetle, *Ips typographus* (Coleoptera: Scolytidae) in the laboratory and under field conditions. Biocontrol Sci. Technol. 2004; 14(8):837-848.

21. Lacey LA, Siegel JP. Safety and Ecotoxicology of Entomopathogenic Bacteria, in Entomopatgenic Bacteria: From Laboratory to Field Application, 2000, 144.

22. Lacey LA, Frutos R, Kaya HK, Vail P. Insect pathogens as biological control agents: Do they have a future. Biological Control. 2001; 21:230-248.

23. Meadows MP. Bacillus thuringiensis in the environment-ecology and risk as-sessment. In: Entwistle PF, Cory JS, Bailey MJ, Higgs S. eds. *Bacillus thurin-giensis*: an environmental biopesticide; theory and practice. Chichester, John Wiley, USA, 1993, 193-220.

24. Mendonsa ES, Vartak PH, Rao JU, Deshpande MV. An enzyme from *Myrothecium verrucaria* that degrades insect cuticles for biocontrol of *Aedes aegypti* mosquito. Biotechnol. Lett. 1996; 18(4):373-376.

25. Microbial control of pests and plant diseases. London, Academic Press Inc. 738-768.

26. Moscardi F. Assessment of the application of baculoviruses for control of Lepidoptera, Annual Review of Entomology. 1999; 44:257-289.

27. Nagrare VS, Kranthi S, Kranthi KR, Chinna Babu Naik, Rishi Kumar B, Dharajothi SS. Cotton Plant Health Manual Central Institute for Cotton Research, Nagpur, 2013.

28. Nicholson GM. Fighting the global pest problem: Preface to the special Toxi con issue on insecticidal toxins and their potential for insect pest control, Toxicon. 2007; 49:413-422.

29. Salunkhe R, Patil CD, Salunkhe BK, Rosas-Garcı´a NM, Patil SV. Effect of wax degrading bacteria on life cycle of the pink hibiscus mealybug, Maconellicoccus hirsutus (Green) (Hemiptera: Pseudococcidae). Bio Control. 2013; 58:535-542.

30. Shaikh SA, Deshpande MV. Chitinolytic enzymes: their contribution to basic and applied research. World J Microb. Biot. 1993; 9(4):468-475.

31. Thakore Y. The biopesticide market for global use. Industrial Biotechnology. 2006; 2:194-208.

32. Watson GW, Kubiriba J. Identification of mealybugs (Hemiptera: Pseudococcidae) on banana and plantain in Africa. African Entomol. 2005; 13:35-47.

Chapter - 5
Eco-Friendly Management of Plant Parasitic Nematodes-Recent Trends and Futuristic Approach

Authors

Sankari Meena K
ICAR-National Rice Research Institute, Cuttack, Odisha, India

Boopathi T
ICAR-Research Complex for NEH Region, Mizoram Centre, Kolasib, India

Arunkumar N
Department of Agricultural Microbiology, Tamil Nadu Agricultural University, Coimbatore, Tamil Nadu, India

Guru Prasanna Pandi G
ICAR- National Rice Research Institute, Cuttack, Odisha, India

Mayabini Jena
ICAR- National Rice Research Institute, Cuttack, Odisha, India

Chapter - 5

Eco-Friendly Management of Plant Parasitic Nematodes-Recent Trends and Futuristic Approach

Sankari Meena K, Boopathi T, Arunkumar N, Guru Prasanna Pandi G and Mayabini Jena

Introduction

Nematodes are complex group of microorganisms which can survive well in the entire environment. They are microscopic, pseudocoelamate, unsegmented, vermiform with transparent cuticle. The name nematode was derived from the Greek words Nema (thread) and oides (resembling). Nematodes are called by different names in different parts of the world. They are called as eelworms in Europe, Nemas in the United States and round worms by zoologists. Nematodes thrive well in different types of habitats from Deep Ocean to mountain tops, from hot water springs to antartic regions, barren lands to cultivated fields. They occupy each and every habitat on the earth. Soil in a hectare of agricultural ecosystem contains about billions of plant parasitic as well as free living nematodes.

The female nematodes of some plant parasitic nematodes become swollen, pyriform or saccate upon maturity. Size of the nematode may vary from few 100 microns (plant parasitic form) to few meters (animal parasitic form). Nematodes posses all the systems of higher animals except the system of circulation and respiration. Plant feeding types possess a needle like Protrusible stylet or spear which is used as a feeding apparatus. In male nematodes, reproductive system opens directly into the rectum forming a cloaca while the reproductive system of female nematodes opens as a separate opening. Vulva generally located on the ventral side of nematode body.

Role of plant parasitic nematodes in agricultural production has remained unnoticed due to their soil borne nature, microscope size and hidden mode of life (either in the soil or inside the plant tissues). The symptoms caused due to nematode damage are non typical which make it difficult for the agriculturists to judge the plant health on the basis of appearance of symptoms on above ground plant parts. Damage caused by the nematodes in plants is either direct or indirect. The direct damage is generally due to feeding by the nematodes or extraction of food from the host with the help of stylet and mechanical destruction of the host cells. The disease spreads gradually in the plants which lead to decline in crop yields. Nematode infection may not kill the plants but make them to become weak and leads to loss of vigour thereby making them susceptible to other diseases causing organisms and other abiotic stresses such as heat, drought and etc., Extent of damage in plants is directly proportional to the prevailing nematode population in the field.

In many legume crops, nematode infestation interferes with nodulation and adversely affects nitrogen fixation. It has also been seen that nematode infestation during early stages of germination may sometimes leads to pre emergence death of seedlings or poor plant growth and making the plants unable to recover from the initial set back received due to nematode infection. Indirect damage due to nematodes got manifested as a result of interaction between nematodes with bacteria, fungi, viruses and other soil borne cohabiting microbes. Hence, crop losses due to combined effects may be additive and synergistic and much more than the total individual effects.

Important Plant Parasitic Nematodes

Plant parasitic nematodes that cause major loss to the crops are listed below.

i) Seed Gall Nematode (*Anguina tritici*)

It is important nematode infesting wheat in northern regions of India. It causes ear cockle disease in wheat. The nematodes along with the bacteria, *Corynebacterium tritici* cause tundu disease or yellow slime. Overall damage of wheat crop due to this nematode during severe infestation was reported to be about 80 per cent.

ii) Cereal Cyst Nematode (*Heterodera avenae*)

It is an important nematode infesting wheat and barley in Rajasthan, Punjab, Haryana, Himachal Pradesh and Jammu and Kashmir. Attack of this nematode in wheat and barley results in molya disease.

iii) Root Knot Nematode (*Meloidogyne* spp.)

It is a polyphagous nematode infesting many crops like vegetables, fruits, pulses and ornamentals. Infestation of this nematode results in the formation of root knot or root galls.

iv) Potato Cyst Nematodes (*Globodera* and *Heterodera* spp.)

Two species of potato cyst nematodes namely, *Globodera rostochiensis* and *G. pallida* are considered to be the major nematode problem associated with the potato crops cultivated in Nilgiris district of Tamil Nadu. Severe infestation of this nematode results in total failure of the crop.

v) Citrus Nematodes (*Tylenchulus semipenetrans*)

This nematode is associated with the slow decline disease of citrus. It is also responsible for the die back disease of citrus in India.

vi) Reniform Nematode (*Rotylenchulus reniformis*)

It is considered to be one among the important nematodes infesting cotton, maize, cow pea and black gram.

vii) Root Lesion Nematode (*Pratylenchus coffeae*)

It is an important nematode infesting coffee in South India.

viii) a) Burrowing Nematode (*Radopholus similis*)

It cause

a. Pepper yellows in Indonesia

b. Banana rhizomes rot in various parts of the world.

c. Slow wilt of pepper in Karnataka

b) Burrowing Nematode (*Radopholus citrophilus*)

It cause spreading decline of citrus in Florida (Jonathan, 2010).

Management of Plant Parasitic Nematodes

Plant parasitic nematodes are one of the major pathogens infesting the crop plants worldwide and limiting the production of the crops. Control of nematodes is a difficult task, since they are subterranean pests and microscopic in nature Neamtode control mainly relied on chemical nematicides for more than 2 decades since remarkable reduction of nematode population has been achieved using the chemical nematicides. Though nematicides are effective in managing the nematodes, they are unaffordable to the small farmers and their application have detrimental effect towards the environment.

Ecofriendly and economically viable alternative methods to chemical nematicides were found to be effective in nematode management since they aim to improve the quality and yield of the crop by keeping the nematode population below the economic threshold level. These methods are infact a profitable means of nematode management.

Ecofriendly methods generally followed in nematode management include:

I. Physical Control
II. Cultural Control
III. Biological Control
IV. Host Plant Resistance

I. Physical Control

Physical control includes physical cleaning of infested seed, application of heat (Direct heat, hot water treatment, solarization and sterilization), mechanical forces, irradiation and osmotic pressure and etc. However, this method is more difficult to employ for large quantities of soil and plant materials. These physical treatments may cause hazardous effects to plants or operators if not carried out properly. The most important physical methods of nematode control are the application of energy as heat to produce a temperature exceeding the limit of tolerance for the nematode. Direct heating of soil at 55 – 60°C for 30 minutes eliminate plant parasitic nematodes. Sterilization of soil by steam in greenhouse beds, seedbeds and other small areas can control insects, weed seeds, nematodes, bacteria and fungi. The most important nematodes controlled by greenhouse steaming are root knot nematode on tomato, cucumbers (Sterilizing soil using steam at 70°C for 30 minutes destroy most PPNs).

Similarly, heating of upper field soil by spreading waste material over field and putting fire is the primitive way of heating soil. (e.g.) Burning of 24 kg of rice husk/m^2 on tobacco seed bed and putting fire opposite to wind current for prolong heating effect in tobacco, it frees soil nematode and this process is called as "Rabbing". The left out straw/stubble in the field after harvest may be effective in protecting the next crop from nematode attack. Soil can be disinfested at the temperature of 40-45°C for short period of time to eliminate the harmful nematodes. Soil solarization is an important method of nematode management. It can be done by covering the moistened soil with transparent polythene sheet (8 μm) for about a month during the hot season. In this method, the primary source of energy for heating the soils is composed of short wave of (0.24 μm) solar radiation and long wave (4 – 8 μm) of atmosphere radiation. Increase of temperature in wet plastic mulched soil is primarily due to the elimination of heat loss by evaporation and heat conviction during the daytime and partially due to the greenhouse effect i.e. the prevention of part of the long wave radiation from leaving the ground.

The latter effect is due to the formation of droplets on the inner surface of the plastic film reducing its transmissivity.

Hot water treatment of planting material is widely used for the control / killing of nematodes within tissues before planting / sowing and has proved useful with nematode infested bulbs, corms, tubers, seed and roots of plants such as chrysanthemum, strawberry, banana, rice etc. This method requires immersing of nematode infested material into thermostatistically in a controlled waterbath at different temperature for different length of time.

E.g. Potato - 45°C for 5-10 min

Rice - *A. besseyi* - 50-57°C for 5-15 min

Wheat - *A. tritici* - 54-55°C for 10-12 min.

Desiccation of soil for a shorter period of time of one month or less causes 80% mortality of nematodes in case of susceptible species (e.g. *Ditylenchus dipsaci*) but for more tolerant species more than six months may be needed (eg. *Rotylenchulus reniformis*)

In arid climate a field population of RKNs can be reduced by ploughing at intervals of 2-4 weeks during the dry season. This exposes eggs and juveniles present in root and deeper layers of the soil to rapid desiccation. Rewetting the soil for a shorter time during the period of desiccation is expected to improve nematode control owing to the increased susceptibility of reactivated nematodes to desiccation.

Soft bodied, normal turgid nematode bodily structure can be easily injured by mechanical forces exerted upon it. Nematode in coarse textured soil can be injured by abrasion caused by agitating the soil intensively. Irradiation is another mechanism of physical force to kills the nematode. Cysts of *G. rostochiensis* exposed to 20,000 γ were found to containonly dead eggs and at 40, 000 γ exposure, the eggs lost their contents. About 100% nematode mortality was observed when sucrose or dextrose were added to nematode infested soil @ 1 to 5% by weight (Feder, 1960). But these methods are not practical and economical. Seed washing, seed cleaning and winnowing are the effective physical methods to remove the seed galls from normal healthy wheat seeds. Application of ultrasonic have little effect on *Heterodera* spp. but the use of this ultrasonics is not practically feasible (Jonathan, 2010).

II. Cultural Control

Normally, the cultural practices will be carried out at little or no extra expense. Crop rotation attempts to keep nematode population to a level at which crop damage is reduced to a minimum. By growing non-host crops, the numbers of plant parasitic nematodes can be reduced. Monocropping with potatoes increased potato cyst nematode. Crop rotation with non host crops *viz*., wheat, strawberry, cabbage, cauliflower, peas, maize and beans for more than 3-year reduce the nematode population to a safe level. Fallowing is an alternative method to crop rotation wherein the field is left without cultivation, preferably after ploughing which helps to expose the nematodes to sunlight and they will die due to starvation without host plant. This method is not economical. In plants, propagated by vegetative means we can eliminate nematodes by selecting the vegetative part from healthy plants. The golden nematode of potato, the burrowing, spiral and lesion nematodes of banana can be eliminated by selecting nematode free planting materials. The wheat seed gall nematode and rice white tip nematode can be controlled by using nematode free seeds.

During the onset of summer, the infested field is ploughed with disc plough and exposed to hot sun, which in turn enhances the soil temperature and kills the nematodes. Green manures and green leaf manures have their own effect in the control of nematodes. Raising green manure crops and addition of more amount of farm yard manure, oil cakes of neem and castor, pressmud and poultry manure etc enriches the soil and further encourages the development of predacious nematodes like *Mononchus* spp. and also other nematode antagonistic microbes in the soil which checks the parasitic nematodes in the filed.

Flooding is an important practice to control nematodes affecting vegetables. Keeping the land submerged in water decreased O_2 content of the soil and kills the nematodes by asphyxiation. Chemicals lethal to nematodes such as butyric and propionic acid, H_2S and NH_3 often develop in flooded soil. Nematode life cycle depends on the climatic factors. Adjusting the time of planting will help to avoid the nematode damage. In some cases crops may be planted in winter when soil temperature is low and at that time the nematodes cannot be active at low temperature. Early potatoes and sugar beets grow in soil during cold season can escapes cyst nematode damage since the nematodes are not that much active, to cause damage to the crop during cold season.

Growing of trap crops, antagonistic crops and cover crops plays a major role in the reduction of nematodes in the field. Two crops are grown in the field, out of which one crop is highly susceptible to the nematode. The nematode attacks the susceptible crop. By carefully planning, the susceptible crop can be grown first and then removed and burnt. Thus the main crop escapes from the nematode damage. Eg. Cowpea is highly susceptible crop can be grown first and then removed and burnt. Cowpea is highly susceptible to root knot nematode and the crop can be destroyed before the nematodes mature.

The root leachates / root chemical / allelochemical of certain plants are toxic to nematodes and growing such crops will bring down significant reduction in nematode population. In Marigold (*Tagetes* spp.) plants, α – terthinyl and bithinyl compounds are present throughout the plant from root to shoot tips are found effective against root knot and burrowing nematodes. Pangola grass produce pyrocatechol which are effective against *Meloidogyne incognita*, *Meloidogyne javanica*, *Meloidogyne hapla* and *Meloidogyne arenaria*. Asparagus produce Asparagusic acid against *T. chiristei.*

Cover crops like *Calapogonium* and green manure crops like sunnhemp and daincha and green leaf manure like neem, *water hyacinth* and pungam were found to be effective in the control of nematodes. Nematode resistant varieties have been reported from time to time in different crops. Use of resistant varieties is a very effective method to avoid nematode damage. Nemared, Nematex, Hisar Lalit and Atkinson are tomato varieties resistant to *M. incogntia*. Potato variety Kufri swarna is resistant to *G. Rostochiensis* (Jonathan, 2010).

III. Biological Control

Biological control of plant parasitic nematodes is accepted as a key practice in sustainable agriculture, since it is based on the management of a natural resource, i.e., certain rhizosphere organisms. Thus biological control preserves environmental quality by reduction in chemical inputs (Barea and Jeffries, 1995). Plant parasitic nematodes have evolved a protective structures and adaptions for their survival. Body of the nematode is protected by a multi layer cuticle which acts as a barrier for adverse environmental conditions. Eggs of the nematodes *viz.*, *Meloidogyne, Tylenchulus* and *Rotylenchulus* are laid in egg masses in gelatinous matrix. In *Heterodera*, eggs are retained within the body of the female which acts as a protective layer. Behavior modifications occur in the anhydrobiotic states reduce the possibility of the nematodes by predation and parasitism. High reproductive capacity and shorter duration of life cycle of the nematodes makes them difficult to control. Endoparasite, ectoparasite and above ground feeders have different mode of feeding habits and that should be considered for devising biocontrol strategies when concerning about a particular nematode genera.

Sedentary Endo Parasites

Sedentary endoparasites are difficult target for microbial antagonistic organisms. Since *Meloidogyne* females lay eggs outside the root surface, eggs appeared to be vulnerable to the attack by antagonists. In certains cases, where the eggs are laid galled tissues (eg. *Meloidogyne graminicola*) in rice. In such cases, antagonists are unable to exert effective control over the eggs.

Semi Endoparasites

In this group, (*Tylenchulus, Rotylenchulus*) juveniles feed ectoparacitically and the saccate female are partly observed on the root surface throughout their lives and eggs are protected by the gelatinous matrix. Nematode trapping fungi (Walter and Kablon, 1990) and egg parasitic fungi, *P. lillacinus* were effectively reduce the population of citrus and reniform nematodes.

Migratory Endoparasites

Migratory endoparasites (*Pratylenchus, Radopholus*) are difficult to control since they spend most of the life within the roots and found in the soil only when the host plant is diseased. They lay their eggs within the root tissues. Juveniles hatch and develop without moving out from the roots.

Ecto Parasites

Although they do not receive any protection from the roots, their migratory nature makes them difficult to control by the soil antagonists.

Above Ground Feeders

Nematode genera such as *Angunia, Ditylenchus* and *Aphelenchoidus* feed on stem, leaves and flower tissues. Efforts should be made to develop suitable biocontrol strategies for above ground feeders (Cayrol and Frankowsi, 1986).

Biocontrol of Nematodes

Biocontrol of nematodes with microbial agents is now gaining interest among nematologists because it is environment friendly and economically feasible method. It does not allow the nematodes to develop into new races or biotypes (Linford and Yap, 1939). Biological control results from the action of soil microorganisms and the soil microfauna and is mediated through mechanisms such as parasitism, predation, competition and antibiosis (Stirling, 1991). There are several microbial pathogens which are effective against plant parasitic nematodes. They include the obligate bacterial parasite, *Pasteuria penetrans* and Plant growth Promoting

Rhizobacteria (*Bacillus subtilis, Pseudomonas fluorescens, Burkholderia cepacia* etc.,). Nematicidal fungi include predacious or trapping fungi, endoparasitic fungi, fungal pathogen/parasites of females, endomycorrhizal and mutualistic fungi. They include *Trichoderma harzianum, Hirsutella rhossiliensis, Hirsutella minnesotensis, Verticillium chlamydosporum, Paceilomyces lilacinus* and *Myrothecium verrucaria.* Entomopathogenic nematode, *Steinernema riobravis* can provide control against root knot nematode (Vyas *et al.*, 2009). Although the exact mechanism of control is not known, researchers hypothesize that the allelochemical produced by *S. riobravis* repels plant parasitic nematodes.

Principles of Biological Control

Usually biological controls operate more slowly but are generally more stable and long lasting than chemical and cultural measures because different methods are effective at different times and locations under varying climatic conditions. Various approaches followed in the biological control are listed below.

Biocontrol can be categorised into Two Types

a) Biological Control with Resident Antagonists

Suppressive soils are soils in which the pathogens establish but cause no damage or cause some disease damage, but the disease becomes progressively less severe even though the pathogen persists in soil (Cook and Baker, 1983). Most soils suppress nematodes to some degree. Suppression is directly related to the total amount of microbial activity in the soil or plant at a critical time in the life cycle of the nematodes. Initial objective of biological control is to maximize soil suppressiveness through the manipulation of resident antagonists. These microorganisms probably resided in equilibrium with the nematodes and plants before the intervention of agriculture.

b) Biological Control with Introduced Antagonists

This can be achieved through,

a) Root colonization process

b) Seed treatment

c) Composites of antagonists

To achieve protection against nematodes, biocontrol organism needs to be rhizosphere competent (Ahmad and Baker, 1987). Certain mutants of the fungus, *Trichoderma* produce more cellulose than their wild type parent strains which were found to be rhizosphere competent. The effectiveness of introduced antagonists for biological control depends on survival, establishment or proliferation in the infection site on the root. The microbiological balance is altered by various environmental factors such as heat treatments, nematicides and other chemical treatments that eliminate groups of microorganisms. Such treatment favours the introduced strains of microbial antagonists to plant pathogens and nematodes. Introduced antagonists have a greater tolerance of broad-spectrum biocides than many other soil microorganisms and colonize treated soils more rapidly than other soil competitors.

Composites of Antagonists in the Biocontrol

Mixtures of biocontrol agents are particularly effective since they have different modes of action or different ecological attributes, so that they function at different times or locations. This approach mainly requires compatibility among the antagonists. Composites of microorganisms can control several plant diseases. Combinations of fluorescent pseudomonads and *Trichoderma species* can effectively control most of the soil borne pathogens and nematodes.

Advantages of Biocontrol Agents

- Less expensive compare to chemicals
- Easy marketability
- Ecofriendly
- No resurgence problems
- Once establish in the field, they remain viable for a longer period

Disadvantages of Biocontrol Agents

- Lack of awareness
- Lack of standardize technics for mass production
- Species specificity
- Variation in the performance due to interaction with other soil borne pathogens
- Non compatibility with other management practices

IV. Host Plant Resistance

Host plant resistance is an environment friendly and economically viable method.

There involved two types of resistance

1. Passive resistance

2. Active resistance

Passive or pre infectioal resistance is mainly due to physiological and chemical barriers present in the plant which restrict the entry of nematode into the plants. It involves the production of toxins by the plants (eg. Production of several compounds like i. glycoside compounds in *Asparagus officinalis* against *Trichodorus christiei*; ii. α-terthienyl compounds in Marigold against *Meloidogyne* spp. and *Pratylenchus* spp. (Winoto, 1969) iii. Pyrocatechol in *Eragrostis curvula* against *Meloidogyne* spp. IV. Terpenoid aldehydes in cotton against *Meloidogyne incognita*).

Active or post infectional resistance is based on the hypersensitivity of the plants to the nematode infestation. The host-parasite interaction stimulates the biochemical reactions in the host plants which lead to necrotic cells. This necrotic cells make the nematodes to deprive of their foods source which delays the development of nematodes and make them to die (Wallace, 1961). HPR is one of the emerging fields in nematode management. Breeding of crop plants against nematodes will make the plant resistant to nematodes. It does not need any specialized applications which will reduce the input cost. Moreover, it is an ecofriendly method of nematode management. Understanding the genetic diversity of nematodes is the important factor to be considered for developing resistant source against nematodes. The techniques such as somatic hybridization, embryo rescue and direct gene transfer should be enhanced to get efficient transfer of genes (Giebel, 1982).

Plant resistance has been identified and incorporated for the nematodes like *Meloidogyne, Heterodera, Rotylenchulus, Tylenchulus*, and *Ditylenchus*. Host plant resistance was identified against *Meloidogyne* species in crops like beans, carrot, cowpea, cotton, soybean, sweet potato, tobacco, grape and tomato. Similary, HPR against *Globodera species was identified in potato and tobacco. Resistant citrus plants were recorded against Tylenchulus semipenetrans* (Roberts, 1982).

Biochemical and Molecular Mechanism of Resistance against Nematodes

Biochemical basis of resistance against root knot nematode was identified in tomato and cowpea. Based on the activities of the enzymes *viz.*, Peroxidase (PO), Indole acetic acid (IAA), Phenylalanine ammonia lyase

(PAL), Thyrosine ammonia lyase (TAL), Poly phenol oxidase (PPO), Superoxide dismutase (SOD) and acid phosphate upon the inoculation of the nematodes, reaction of the host to that particular nematode was calculated. Based on these reports, a model was proposed to study the plant - nematode interaction (Premachandran and Dasgupta, 1983).

Various hydrolytic enzymes *viz*., amylase, esterases, trehalose, cellulose and proteases were detected from the infective stage of the root-knot nematode, *M. incognita*. Presence of these hydrolytic enzymes helps the infective juvenile to hydrolyse the host components like carbohydrates and proteins and alters the host metabolism which ultimately leads to the diseased condition of the hosts (Ganguly and Rao, 2009). Microscopic nature of the plant parasitic nematodes makes the identification of the nematode genera a difficult task through the taxonomic studies. Accurate identification is very important to provide the effective management strategy against nematodes. Molecular assays based on protein and DNA was widely used in recent days for the identification and description of the nematode species. Esterase enzyme pattern is commonly used to identify *Meloidogyne* spp. (Gangulyand Dasgupta, 1989), *Heterodera* spp., *Rotylenchulus* spp. and etc. In the esterase band pattern, intraspecific variation were observed in *the nematodes tested.* In DNA based diagnostics, variation can be expressed in terms of a particular gene. In PCR, millions of targeted DNA sequence can be obtained and they can be further characterized by RFLP, AFLP, dot blotting or sequencing.

RNA Interference in Nematode Management

RNAi is being used in recent times in the field of nematode management. In the process of RNAi, the RNA molecules inhibit the expression of genes thereby elucidate the functions of the particular gene. This technology provides a promising future in nematode management. In this method, Plants can be designed to produce dsRNAs to silent the specific gene responsible for the parasitism and growth of the nematode. Upon feeding on these plants, the nematode will ingest those dsRNA from the plant cytoplasm which would inactivate the gene targeted by the dsRNA and stop the further parasitism and growth of the particular nematodes (Ganguly and Rao, 2009).

RNAi has effectively been applied in many nematodes. In *Heterodera glycines* and *Globodera pallida,* RNAi was applied by targeting on Cystein-proteinase of the nematode which resulted in changing the sex ratio of the nematode from female to male in the ratio of 3:1 to 1:1 (Urwin *et al*. 2002). The study conducted by Rosso *et al*. (2005) in *Meloidogyne incognita* revealed the knock down of two genes in subventral gland of the nematode. Transgenic tobacco targeted on eggs and eggs producing females of *Meloidogyne javanica* recorded drastic changes in the fecundity and development of the nematode (Fairbairn *et al.*, 2007).

Conclusion

Impact of climate change coupled with irrelevant land use pattern and modern agricultural practices will pose a new threat to the agricultural production in terms of biotic stresses created by plant parasitic nematodes. Inappropriate use of chemical nematicides is posing a serious threat to the environment. Incorporation of ecofiendly methods in appropriate time against the nematodes will be an effective alternative to the synthetic nematicides on the long run. Use of molecular techniques will be a reliable and accurate method for nematode identification which can help in the development of suitable management strategies against nematodes. However, appropriate standardization of these techniques should be done by testing the efficacy of the products under natural field condition. Hence, in a long run, the ecofriendly approaches will be more beneficial over the chemical nematicides and they will reduce the frustration faced by the farmers due to high cost of nematicides.

References

1. Ahmad JS, Baker R. Rhizosphere competence of *Trichoderma harzianum*. Phytopathol. 1987; 77(2):182-189.
2. Barea JM, Jeffries P. Arbuscular mycorhizas in sustainable soil plant systems. In: Mycorrhizae structure, function, molecular biology and biotechnology B. Hock and A. Varma (Eds.). Springer, Heidelberg, 1995, 521-559.
3. Cayrol JC, Frankowsi JP. Influence of number of parasitizing conidia of *Hirsutella rhossilensis* on the mortality of *Ditylenchus dipsaci*. Rev. Nematol. 1986; 9:411-412.
4. Cook RJ, Baker KF. The nature and practice of biological control of plant pathogens. APS Press, St. Paul, MN, 1983.
5. Fairbairn DJ, Cavallaro AS, Bernard M, Mahalinga Iyer J, Graham MW, Botella JR. Host-delivered RNAi: an effective strategy to silence genes in plant parasitic nematodes. Planta. 2007; 226(6):1525-1533.
6. Ganguly AK, Rao U. Decades of researches in biochemical and molecular Nematology at IARI. J Plant Protec. Sci. 2009; 1(1):1-8.

7. Giebel J. Mechanism of resistance to plant nematodes Ann. Rev. Phytopathol. 1982; 20:257-279.

8. Jonathan EI. In: Nematology-Fundamentals and applications. New India Publishing Agency, New Delhi, 2010, 7-285.

9. Linford Yap. Root knot nematode injury restricted by a fungus. Phytopathol. 1939; 29:596-608.

10. Premachandran D, Dasgupta DR. A theoretical model for plant nematode interaction. Review de Nematol. 1983; 6:311-314.

11. Roberts PA. Plant resistance in nematode pest management. Journal of Nematol. 1982; 14:24-33.

12. Ross MN, Dubrana MP, Clmbotini N, Jaubert S, Abad P. Application of RNA Interference to root knot; Nematode genes encoding oesophageal gland proteins. Mol.Plant-Microbe Interact. 2005; 18(17):615-620.

13. Stirling GR. Biological control of plant parasitic nematodes. CAB International, Wallingford, UK, 1991.

14. Urwin PE, Lilley CJ, Atkinson HJ. Ingestion of double stranded RNA by preparasitic juvenile cyst nematodes leads to RNA interference. Mol.Plant-Microbe Interact. 2002; 15:747-752.

15. Vyas RV, Maghodia AB, Biren A, Patel Patel DJ. Field Efficacy of Entomopathogenic Nematode, *Steinernema riobrave* and their toxins against root knot nemtode, Meloidogyne spp. on okra. Indian J Nematol. 2009; 39(2):29-231.

16. Wallace HR. The nature of resistance in chrysanthemum varieties to Aphelenchoides ritzemabosi. Nematologica. 1961; 6:49-58.

17. Winoto SR. Studies on the effect of *Tagetes* species on plant parasitic nematodes. Sticht. Fonds. Landbouw. Export Bur. Publ. 1969; 47:132.

Chapter - 6
Silicate Solubilizing Bacteria-Role in Soil Nutrition and Plant Protection

Authors

Arunkumar N
Department of Agricultural Microbiology, Tamil Nadu Agricultural University, Coimbatore, Tamil Nadu, India

Priyanka S
Department of Plant Breeding and Genetics, Tamil Nadu Agricultural University, Coimbatore, Tamil Nadu, India

Guru Prasanna Pandi G
ICAR- National Rice Research Institute, Cuttack, Odisha, India

Sankari Meena K
ICAR- National Rice Research Institute, Cuttack, Odisha, India

P. Panneerselvam
ICAR- National Rice Research Institute, Cuttack, Odisha, India

T. Boopathi
ICAR - Research Complex for NEH Region, Mizoram Centre, Kolasib, Mizoram, India

Chapter - 6

Silicate Solubilizing Bacteria-Role in Soil Nutrition and Plant Protection

Arunkumar N, Priyanka S, Guru Prasanna Pandi G, Sankari Meena K, P. Panneerselvam and T. Boopathi

1. Introduction

Soil consists of sand, silt, and clay, all of which possesses either primary or secondary form of silicate minerals. Weathering and biological breakdown i.e., a variety of organic weathering agents arising out of decomposition of plant and animal residues, root exudates andor metabolites from rhizosphere and non rhizosphere micro organism might create an environment congenial for the breakdown of silicate minerals releasing the ionic component (Boyle and Voigt, 1973). The breakdown of natural and synthetic silicates by microorganisms was demonstrated by Webley *et al.* (1963). These, mineral's solubilizing cultures similar to biofertilizer can be isolated and identified to release minerals such as silicon, phosphorus and potassium that are abundantly available in the cultivated field itself. Further, there are several wastes obtained from mining and mineral related industries that may be useful for low input agriculture.

Some of the wastes or its by products disposed such as wastes from granite crusher yard and basic slag from steel industry can be reutilized as a potential silicate sources providing silicon nutrition for agricultural purposes. Jeffery *et al.* (2006) reported that silicate incorporation in soil reduced the incidence of Damping-off (*Pythium ultimum*) of cucumber when silicate was incorporated in potting mixture. There appeared to be a minimum time for accumulation of silicon by seedlings, to obtain protection from challenges by pathogens. Hence knowledge on soil microorganisms capable of silicate solubilization and their role in pest and disease incidence in agricultural field gain importance. Consequently, farmers, farm scientists and policy planners search for ecofriendly and environmentally safe on-farm inputs to minimize environmentally destructive and economically disruptive off farm inorganic inputs so that the resources available today are passed on to the posterity without exhaustion or loss in productivity.

2. Silicon

Silicon is an element abundantly available in earth's crust. It is second only to oxygen (Ehrlich, 1981). Few organisms absorb and accumulate silicon. The most notable once are diatoms, chrysophytes, silicoflagellates, some xanthophytes, radioarians, plants, insects and vertebrates (Heinen, 1962). Plants take up different quantities of silicon according to their species (Russell, 1973). The silica accumulators were found in the monocotyledonous plant family's Graminae, Cyperaceae and Musacaeae. Monocotyledonous plants accumulate more quantities of silica than dicotyledons. Among the plants, silica concentrations are found to be higher in monocotyledons than in dicotyledons and its level increased from legumes < fruit crops, vegetables < grasses < grain crops (Thiagalingam *et al.*, 1977). Grasses accumulate 2 to 20 per cent Si of the foliage dry weight as hydrated polymer or silica gel. Rice accumulates 4-20 per cent silicon in straw and almost every part of rice contains this element which is not at all added exogenously as fertilizer as done with nitrogen, phosphorus and potassium the trinity of nutrients. Silica depletion can occur in traditional rice soils from the continuous monoculturing of high yielding cultivars with intensive cultivation practices (Miyake, 1993). Therefore silicon management may be important for sustaining rice productivity (Correa Victoria *et al.*, 1994).

2.1 Silicon in Soil

Silicon is abundant in the earth's crust (constituent's upto 28%) and occurs in various complex mineral forms. The main forms of silicon in soil are dissolved silica mainly monosilicic acid (H_4SiO_4) and polysilicic acid ($SiO_2.nH_2O$) (McKeague and Cline, 1963), crystalline and amorphous silica, Si $(OH)_4$ and Mn silicates (Ponnamperuma, 1972). Colloidal forms of silica exist at high temperatures or at saturation levels only formed by acid conditions (Hall, 1972). Sandy and sandy loam soils contain more of total silica and less available and water soluble silica. Jones and Handreck (1963) and Mckeague and Cline (1963) also reported that free sesquioxides removed silica from solution by absorption.

2.2 Silicon in Plants

Silicon (Si) in a chemically combined form is ubiquitous in nature. The Si content of soils can vary dramatically from <1 to 45 % dry weight (Sommer *et al.* 2006), and its presence in the form of silicic acid [$Si(OH)_4$] or its ionized form, $Si(OH)_3O^-$, which predominates at pH > 9 allows its uptake by plants. Silicic acid is generally found in soils at concentrations ranging from 0·1 to 0·6 mM (Epstein, 1994). Si is found in plants at concentrations ranging from 0·1 to 10 % on dry weight basis, an amount equivalent to, or even exceeding, several macronutrients. Plants deprived of Si are often weaker structurally and more prone to abnormalities of growth, development and reproduction and it is the only nutrient which is not detrimental when absorbed in excess (Epstein, 1999).

Silicon does not form a constituent of cellular components because Si - Si bonds are considerably weaker than C - C bonds. Silicon is primarily deposited on the walls of epidermis and vascular tissues conferring strength, rigidity and resistance to pests and diseases. Though silicon has no biochemical role in development of plants, silicon enzyme complexes are found to protect or regulate photosynthesis. Silicon was found to suppress the activity of certain enzymes and the suppression of invertase particularly resulted in phosphatase, which provided a greater supply of essential high energy precursors needed for optimum growth. It was also suggested that silica in plants filter harmful ultraviolet radiation reaching leaf surface with leaf cells acting as 'windows' transmitting the light energy to photosynthetic mesophyll and cortical tissues beneath epidermis, than that would occur if silica were absent. (Tisdale *et al.*, 1985).

Silicon was found to suppress the activity of certain enzymes and the suppression of invertase particularly resulted in greater sucrose production in sugarcane and the reduction of phosphatase provided a greater supply of essential high energy precursors, needed for optimum growth. It was also suggested that the silicon in plants filter harmful ultraviolet radiation reaching leaf surface with leaf cells acting as 'windows' transmitting the light energy to photosynthetic mesophylls and cortical tissues beneath the epidermis, than that would occur if silica were totally absent (Tisdale *et al.*, 1985). Gallo et *al.* (1974) observed that in rice, oat, rye and wheat, seed coat accumulated the most silica and the grain the least. They also observed that leaves and stems of maize and sorghum and leaves of sugarcane and bamboo had highest levels than other plant parts. Yoshida (1975) reported that rice plant absorbs silicon in the form of *ortho*-silicic acid (H_4SiO_4 or $Si(OH)_4$) along with water from the growing medium and the water is lost through transpiration, the silica concentration increases and at higher levels *ortho*-silicic acid polymerizes into silica gel ($SiO_2.nH_2O$) through non enzymatic reactions.

3. Influence of Silicon on Plant Growth

The application of silicon generally increased the uptake of phosphorus, calcium magnesium and the formation of carbohydrate and decreased the uptake of nitrogen, potassium, iron and manganese in transplanted rice

(Islam and Saha, 1969). The application of silicates not only augmented its absorption by rice plant but also had a significant inter relationship with the other nutrients. Silica nutrition has direct and indirect beneficial effects on the rice growth largely due to its unique physiological role (Okuda and Takahashi, 1965; Yoshida, 1975). Application of silica to rice was found to increase the grain yield under both upland and waterlogged conditions (Datta and Shinde, 1985). Addition of 100 mg SiO_2 kg^{-1} to soil increased plant height, tiller numbers, number of leaves, leaf area, dry matter production and leaf area ratio in rice. The transpiration rates of silica supplied plants were reduced. Supply of silica also altered the levels of carbohydrate, protein and phenol and these changes varied slightly with stage of the rice crop and plant part.

Silicon application significantly increased the grain yield by increasing the number of panicles per plant, spicklets per panicle, mature grains per panicle, test weight and harvest index (Ashoka Rani *et al.,* 1997); silica promoted the photosynthetic activity and P utilization in rice (Ma and Takahashi, 1990); water use efficiency of the plants (Vandervorn, 1980). Application of silicate materials increased leaf chlorophyll and decreased leaf freckling in sugarcane. Application of silicates @ 15 metric tons ha^{-1} increased leaf chlorophyll by 65 to 78 per cent (Elawad *et al.,* 1982). Boylson *et al.* (1990) reported that in cotton, silicon was necessary for the development of fibers. Silicon is not only important structurally but it also to play a metabolic role in the synthesis of protein, chlorophyll, DNA, RNA, xanthophylls, lipids and thymidylate kinase (Lewin, 1955).

4. Microbial Dissolution of Silicates

Badr *et al.,* (2006) reported that the release of K and P from feldspar and rock phosphate was substantially enhanced in pure culture inoculated with bacterial strains of silicate dissolving bacteria. *Bacillus cereus* had a much higher potential to release K in the pH 6.5-8.0 than other pH. More aerobic condition produced more K and P than a less aerobic one. Soil pot experiments with sorghum plants indicated that the silicate dissolving bacteria could grow in the tested soils and have a potential of minerals dissolution. The mean effect of silicate bacteria on feldspar dissolution was appreciable greater under the tested soils (62 % of total K) rather than under bacterial culture (53 % of total K). Bacterial inoculation combined with K and P bearing minerals gave 48, 65 and 58 per cent increase in dry matter yield of sorghum plants in clay, sandy and calcareous soil respectively compared to non-inoculated soils. The uptake of K by sorghum plants also increased by 71, 110 and 116 per cent while the uptake of P increased by 41,

93 and 79 per cent in the clay, sandy and calcareous soils respectively.

Friedrich *et al.* (1998) reported that the microorganisms influence the solubilization reaction of silica. Among them, highest leaching activity was observed in the case of *Thiobacillus thiooxidans* in kaoline and quartz sand as model substances. Frequent reports confirm that bioleaching of zinc and nickel from silicate using *Aspergillus Niger* cultures in this citric acid generated by fungi could be an important leaching agent acting in the solubilization of calamine and garnierite. These bioleaching process present two advantages as compared to conventional chemical leaching processes: (a) the very low concentrations of organic compounds present in such a situation represent a lower ecological risk; and (b) even with a lower final yield, the economical cost of such process. Both characteristics could facilitate its industrial application.

In nature, silica scarcely occurs as a monomer, but on the contrary mostly present in a more or less polymerized state and unless depolymerized it cannot be absorbed. Naturally occurring silica can be depolymerized and solubilized by biological action (Heinen, 1962). Biodissolution of polymerized silica into monomeric silica has been demonstrated with *Proteus mirabilis* and *Bacillus caldolyticus.* The *Proteus* culture is able to assimilate some of the monomeric silica (Lauwers and Heinen, 1974). The ability of bacteria to depolymerize crystalline silicates has been shown by Webley *et al.* (1960). The effect of chelation in silicate dissolution was shown in laboratory experiments with a soil strain of *Pseudomonas* that produced 2-keto gluconic acid from synthetic silicate of calcium, zinc and magnesium and from the minerals Wollastonite ($CaSiO_3$), Apophyllite ($KCa_4(Si_4O_{10})_2F \cdot 8H_2O$) and Olivine ($(Mg, Fe)_2SiO_4$).

Muntz (1890) was the first to suggest that bacteria are involved in rock decomposition. Gromov (1957) showed that even the surface of mountain rocks were colonized by algae, bacteria and fungi. The siliceous materials in the rock can be attacked through the products of metabolism of microorganisms was also demonstrated (Webley *et al.*, 1963). He also found that synthetic silicates of calcium and zinc were more easily dissolved than magnesium silicates. Wollastonite a naturally occurring calcium silicate was even less soluble. The fungi cause dissolution of silicates through the production of oxalic acid and citric acid in their metabolism of glucose. The fungus *Aspergillus Niger* M. has been shown to release silicate from apophyllite, olivine, saponite, vermiculite and Wollastonite (Henderson and Duff, 1963).

Duff and Webley (1959) reported silicate dissolving action with a gram-negative bacterium, strain D11 which resembled *Erwinia spp* and with *Bacterium herbicola* or with some *Pseudomonas* strains all of which were able to produce 2- keto- gluconate from glucose. Kalaiselvi (1999) reported that higher dissolution of silica from biotite than from muscovite or feldspar by the silicate solubilizing bacteria isolated from the rice husk. The erosional process of concrete sewer pipes can be explained as the result of several biochemical reactions of the calcium silicate- calcium aluminium matrix. Silica release has been noted in the action of *Penicillium simplicissimus* growing in glucose mineral salts medium in the presence of basalt, granite, granodiorite, rhyolite, andesite, periodotite, dunite and quartzite (Silverman and Munoz, 1970). The attack upon quartz, plagioclase and mepheline by *Penicillium notatum* and *Pseudomonas sp* with release of silica has been reported (Kutuzove, 1969). *Bacillus muciliaginosus* var *siliceous* was also reported to degrade silicate minerals (Mel'nilova *et al.,* 1990; Malinovskaya *et al.,* 1990).

Mera and Beveridge (1993) reported the mechanism of silicate binding to the bacterial cell wall in *Bacillus subtilis.* The initial binding was by formation of silicate to amine complexation in the peptidoglycan of cell wall through electrostatic interaction. The action of an alkaline substance in the dissolution of silica from nepheline, plagioclase and quartz by *Sarcina ureae* growing in a peptone urea broth was also reported (Kutuzova, 1969). Purushothaman *et al.* (1974) reported the distribution of silicate dissolving bacteria in marine environment and suggested that they play a role in cycling of silicon in sea water. They found that majority of isolates were found to be *Psuedomonas sp* and the various isolates differed in the rate of solubilisation of insoluble silicates. Soil bacterium *Proteus mirabilis* was found to take up limited amounts of silicon when furnished in the form of silica gel, quartz or sodium silicate (Ehrlich, 1981). Norkina and Pumpyanskaya (1956) reported that the silicate solubilizing bacterium *Bacillus musiliaginosus* sub sp *siliceous* liberated potassium from feldspar and aluminosilicates in vitro cultures.

5. Silicon Management in Crop Production

Agricultural soils are largely composed of silicate minerals and many soils contain an inadequate supply or naturally low in available silica. Most likely the silica content in some regions might be limiting to sustainable rice production. So silicon management becomes important in increasing rice productivity. Savant *et al.* (1999) reported that sugarcane is known to absorb more Si than any other mineral nutrient, accumulating approximately 380 kg ha^{-1} of Si. Sugarcane responses to silicon fertilization have been documented in some areas of the world, and applications on commercial fields are routine in certain areas. The reason for this plant response or yield increase is not fully understood, but several mechanisms have been proposed. Some studies indicate that sugarcane yield responses to silicon may be associated with induced resistance to biotic and abiotic stresses, such as disease and pest resistance, Al, Mn, and Fe toxicity alleviation, increased P availability, reduced lodging, improved leaf and stalk erectness, freeze resistance, and improvement in plant water economy.

Elawad *et al.* (1982) resulted that application of 15 metric tons ha^{-1} of silicates (Florida slag) increased leaf chlorophyll by 78 and 65 per cent, and decreased leaf freckling by 46 and 41 per cent in first and ratoon crops of sugarcane, respectively. It also increased the levels of Si, P, Ca, and Cu, and reduced the levels of N, K, Mg, Fe, Mn, and Zn in the leaf. And also increased soil pH, Si, P, Ca, and Mg. The application of black or grey ash of rice hulls made rice seedlings healthy and strong and increased their biomass (Savant and Sawant, 1995). Talashilkar and Chavan (1996) reported that application of rice hull ash as a source of silicon has resulted in significant increase in grain and straw yields of rice. Lee (1983) found that rice husk ranked first in influencing the availability of silica in soil which was attributed to quick release of soluble silica by rice husk.

Addition of silicate materials *viz.,* rice husk, furnace slag and sodium Meta silicate at 500 ppm SiO_2 level increased the available silica status in soil (Subramanian and Gopalswamy, 1990). Calcium silicate fertilizers which are made from slags have been applied to paddy fields as a silicon source to increase rice production (Kato *et al.,* 1997). Yield increase is due to siliceous slag application @ 1.5 to 3.0 t ha^{-1} continuously to each crop was also observed in rice in field and pot experiments (Lian and Wang, 1983; Masaoka and Takawan, 1987). Silva (1971) reported that the Si application as calcium silicate, sodium silicate or basic slag increased the solubility of adsorbed soil phosphorus, decreased phosphorus fixation, increased cation exchange capacity and decreased aluminum activity.

The application of calcium silicate to highly weathered savanna soils enhanced the response of applied phosphate in wetland rice (Friesen *et al.,* 1998). Vyas and Motiramani (1971) reported that addition of organic matter and silicate increased the available phosphorus from soil. Chinnaswami and Chandrasekaran (1978) reported that the addition of silica as silica gel to the soils of Tamil Nadu released more of phosphorus which was found to be greater in lateritic sandy loam than in clay and non-lateritic sandy loam soil. Ma and Talahashi (1989) reported that rice straw application to the soil is one of the means to supply a sufficient amount of silicon to the rice plant for healthy growth. Silicon released by straw from two sources, from the silicon contained in the straw and from the soil silicon liberated by straw decomposition. Same field experiments showed that straw application increased the silica content of rice shoots. Kamimura *et al.* (1994) reported that a paddy soil derived from glassy volcanic ash was improved by application of rice straw resulted in the higher percentage of ripened grains and increased the yield.

6. Pest and Disease Resistance Development by Silicate Solubilizing Bacteria

Belanger *et al.* (2003) strongly suggested that Si mediated active localized cell defenses against *Blumeria graminis* attack. His histological and ultrastructural analyses revealed that epidermal cells of Si+ plants reacted to *B. graminis* attack with specific defense reactions including papilla formation, production of callose, and release of electron-dense osmiophilic material identified by cytochemical labeling as glycosilated phenolics. Phenolic material not only accumulated along the cell wall but also associated with altered integrity of haustoria in a manner similar to localized phytoalexins as reported from other pathosystems. Silicon appears to contribute to the suppression of *Eldana saccharina* directly through reduced larval growth and feeding damage to the crop, and indirectly by delaying stalk penetration, resulting most likely in increased exposure time of young larvae to natural enemies, adverse climatic factors, or insecticidal control measures that target young larvae (Walker, 2002).

Fawe *et al.* (1998) reported that silicon is involved in the increased resistance of cucumber to powdery mildew (*Pythium ultimum*) by enhancing the antifungal activity of infected leaves. This antifungal activity was attributed to the presence of low molecular weight metabolites, Phytoalexin. Much of the work on possible mechanisms through which Si affects disease susceptibility has been done on cucumber, but the insights gained from these investigations may also apply to other plants. Soluble Si taken up by plants accumulated in the apoplast, particularly in epidermal cell walls (Epstein, 1994; Marschner, 1995; Samuels *et al.*, 1993). This observation has led many investigators to hypothesize that Si inhibits fungal disease by physically inhibiting fungal germ tube penetration of the epidermis (Datnoff *et al.*, 1997; Belanger *et al.*, 1995). However, subsequent investigators have found that only the trichome bases on the cucumber epidermis tend to became silicified which offers resistance (Belanger *et al.*, 1995; Samuels *et al.*, 1993).

7. Conclusion

Although silicon has not been considered to be an essential element for higher plants, yet its beneficial effects have been demonstrated, especially when they are subjected to biotic or abiotic stresses. Supplying silicon to plants ideally fits in with environmental friendly strategies for sustainable crop production. However, its effect in enhancing plant resistance against bacterial pathogens is not limited to silicon accumulators, and has been described in silicon neutral plants. Elaborate studies on various sources of silicon and bioinoculation of silicate solubilizers and their role in plant protection will throw light on the importance of silicon nutrition. Recent progress in understanding the biological role of silicon in plants will be much helpful in increasing crop yield and enhancing bacterial and viral pathogen resistance.

8. References

1. Ashoka Rani Y, Narayanan A, Saraswathi Devi V, Subbaramamma P. The effect of silicon application in growth and yield of rice plants. Ann. Plant. Physiol. 1997; 11(2):123-128.
2. Badr MA, Shafei AM, Sharaf El-Deen SH. The Dissolution of K and P-bearing Minerals by Silicate Dissolving Bacteria and their effect on Sorghum Growth. Research Journal of Agriculture and Biological Sciences. 2006; 2(1):5-11.
3. Belanger RR, Bowen PA, Ehret DL, Menzies JG. Soluble silicon: its role in crop and disease management of greenhouse crops. Plant Disease. 1995; 79(4):329-336.
4. Belanger RM, Smith CM, Corkum LD, Zielinski B. Morphology and histochemistry of the peripheral olfactory organ in the round goby, Neogobius melanostomus (Teleostei: Gobiidae). Journal of Morphology. 2003; 257:62-71.
5. Boyle JR, Voigt GK. Biological weathering of silicate minerals. Implications for tree nutrition and soil genesis. Plant and Soil. 1973; 38:191-201.
6. Boylsone EK, Hebert JJ, Hensarling TP, Browdow JM, Thibodeaux DP. Role of silicon in developing cotton fibres. J Plant Nutr. 1990; 13:131-148.
7. Chinnaswami KN, Chandrasekaran S. Silica status in certain soils of Tamil Nadu. Madras Agric J, 1978; 65:743-746.

8. Correa-Victoria FJ, Datnoff LE, Winslow MD, Okada K, Frisen DK, Sanz IJ *et al.* Silicon deficiency of upland rice on highly weathered savanna soils in Colombia. II. Diseases and grain quality. IX Conferencia. International De arroz para a Americ Latina e para o Caribe. V. Reuniao Nacional de pesquisa de Arroz. Castro's Park Hotel. Brazil, 1994, 21-25.

9. Datnoff LE, Deren CW, Snyder GH. Silicon fertilization for disease management of rice in Florida. Crop Protection. 1997; 16(6):525-531.

10. Datta NP, Shinde JE. Yield and nutrition of rice under upland and waterlogged conditions. Effect of nitrogen, phosphorus and silica. J Indian Soc. Soil Sci. 1985; 33:53-60.

11. Duff RB, Webley DM. 2-keto-gluconic acid as a neutral chelator produced by soil bacteria. Chem. and Ind. 1959; 10:1376-1377.

12. Ehrlich HI. Geomicrobiology. Marcel Dekker Inc. New York, 1981, 393.

13. Elawad SH, Gascho GH, Street JJ. Response of Sugarcane to silicate source and rate. I. Growth and Yield. Agron. J, 1982; 74:481-484.

14. Epstein E. The anomaly of silicon in plant biology. Proceedings of the National Academy of Sciences. USA. 1994; 91:11-17.

15. Epstein E. Silicon. Ann. Rev. Plant physiol. Plant Mol. Biol. 1999; 50:641-644.

16. Fawe A, Abou-Zais M, Menzies JG, Belanger RR. Silicon-mediated accumulation of flavonoid phytoalexins in cucumber. Phytopathology. 1998; 88:396-401.

17. Friedrich S, Platonova NP, Karavaiko GL, Stichel E, Glombitza F. Chemical and microbiological solubilization of silicates. Acta Biotechnologica. 1998; 11(3):187-196.

18. Friesen DK, Thomas RJ, Rivera M, Asakawa N, Bowen W. Nitrogen dynamics under monocultures and crop rotations on a Colombian savanna oxisol. In: Proceedings of the 16thWorld Congress of Soil Science, Symposium 14, AFES, Montpellier, France (CD-ROM), 1998.

19. Gallo JR, Furlani PR, Bataglia OC, Hiroce R. Silicon content in grass and forage crops. Cienciae cultur. 1974; 26:282-293.

20. Gromov BV. The microflora of rocks and primitive soils in some northern regions of the U.S.S.R. Mikrobiologiya. 1957; 26:52-54.

21. Hall FR. Silica Cycle. In: The Encyclopedia of Geochemistry and Environmental sciences. Encyclopedia of Earth sciences series, (ed.) A. R.W. Fairbridge van Nostrand Reinhold, New York. 1972; IV:1082-1088.

22. Heinen W. Silicium Stoffweschel bei Mikro-organismen II Beziehungen zwischen silicate and phosphate stoffweschsel bei Bakterin. Arch. Mickrobiol. 1962; 41:229-246.

23. Henderson MEK, Duff RB. The release of metallic and silicate ions from minerals, rocks and soils by fungal activity. J Soil Sci. 1963; 14:236-246.

24. Islam A, Saha RC. Effects of silicon on the chemical composition of rice plants. Plant and soil. 1969; 30(3):446-458.

25. Jeffery S, Williams E, Whicker C, Khoo C, Summerell B, Offord C *et al.* Some effects of silicate on *Pythium* Damping-off of cucumbers in potting mixes. Acta Horticulturae. 2006; 450:467-473.

26. Jones LHP, Handreck LK. Effect of iron and aluminium oxides on silica in solution in soils. Nature. 1963; 198:852-353.

27. Kalaiselvi P. Influence of silicate solubilizing bacteria on dissolution of silica and potassium in rice soil added with graded levels of potassium and rice residues. M.Sc. (Ag.) Thesis, TNAU, Coimbator, 1999.

28. Kamimura Y, Furue K, Nishizono N. Improvement of paddy soil derived from glassy volcanic ash (shirasu) by successive applications of organic and inorganic amendments. Soil Sci. Pl. Nut. 1994; 40:9-48.

29. Kato N, Owa N. Dissolution of slag fertilizers in a paddy soil and Si uptake by rice plant. Soil Sci. Plant. Nutr. 1997; 43:329-341.

30. Kutusova RS. Release of silica from minerals as a result of microbial activity. Mikrobiologiya. 1969; 38:596-602.

31. Lauwers AM, Heinen W. Bio-degradation and utilization of silica and quartz. Arch. Microbiol. 1974; 95:67-78.

32. Lee TS. Improvement of latosolic paddy fertility. Soils and Fert, 1983, 1-20.

33. Lewin JC. Silicon metabolism in diatoms III. Respiration and silicon uptake in Novicula pelliculosa. J Gen. Physiol. 1955; 39:1-10.

34. Lian S, Wang CH. Effect of long term applications of siliceous slag on paddy yield and chemical properties of soil. J Agrl. Res. China. 1983;

32:187-197.

35. Ma J, Nishimura K, Takahashi E. Effect of silicon on the growth of rice plant at different growth stages. Soil Sci. Plant Nutr. 1989; 35:347-356.

36. Ma JF, Takahashi E. Effect of silicon on the growth and phosphorus uptake of rice plants under different phosphorus upplies. Pl. Soil. 1990; 126:115-119.

37. Malinovskaya IM, Kosenko LV, Votselko SK, Podgorskii VS. Role of *Bacillus mucilaginosus* polynucleotide in degradation of silicate minerals. Mikrobiologiya. 1990; 59:70-78.

38. Marschner H. Nutritional physiology. In H Marschner, Ed, Mineral Nutrition of Higher Plants. Academic Press Limited, London, 1995, 417-426.

39. Masaoka Y, Takawan N. Effects of silicon in warm season forage crops on cell wall digestion by cellulose. J Japanese Soc. Grassland Sci. 1987; 32:417-418.

40. Mckeague JA, Cline MG. Silica in soil. Advan. Agron. 1963; 15:339-396.

41. Mel'nilova EO, Avakyan ZA, Karayko GI, Krutsko VS. Microbial degradation of Beryllium containing minerals. Mikrobiologiya. 1990; 59:63-69.

42. Mera MU, Beveridge TJ. Mechanisms of silicate binding to the bacterial cell wall in Bacillus subtilis. J Bacteriol. 1993; 175(7):1936-1945.

43. Miyake Y. Silica in soil and plants. Sci. Rep. Faculty of Okayama Liniv. 1993; 81:61-79.

44. Muntz A. Surla decomposition des roches ET la formation de la terrarable. Compt. Renmd. Akud. Sci. Paris. 1890; 110:1370-1372.

45. Norkina SP, Pumpyanskaya LV. Certain properties of silicate bacteria. *Dokl.* Akal. Skh. Nauk. 1956; 3:27-31.

46. Okuda A, Takahashi E. The role of silicon. In: The mineral nutrition of the rice plant. John Hopkins press, Baltimore. Maryland. 1965, 126-146.

47. Ponnamperuma FS. The chemistry of submerged soils. Adv. Agron. 1972; 24:29-96.

48. Purushothaman A, Chandramohan D, Natarajan R. Distribution of silicate dissolving bacteria in vellar estuary. Curr. Sci. 1974; 43:282-

283.

49. Russell EW. Soil conditions and plant growth. 10th ed. Longman London, 1973, 756.

50. Samuels AL, Glass ADM, Ehret DL, Menzies JG. The effects of silicon supplementation on cucumber fruit: changes in surface characteristics. Annals of Botany. 1993; 72:433-440.

51. Savant NK, Sawant AS. Nutrition composition of rice seedlings as influenced by rice hull ash application to seedbed. *Oryza,* in press, 1995.

52. Savant NK, Korndörfer GH, Datnof LE, Snyder GH. Silicon nutrition and sugarcane production. Plant Nutr. 1999; 22:1853-1903.

53. Silva JA. Possible mechanism of crop response to silicate application. Proc. Int. Symp. Soil Fert. Evaluation. 1971; 1:805-814.

54. Silverman MP, Munoz EF. Fungal attack on rock solubilization and altered infrared spectra. Science. 1970; 169:985-987.

55. Sommer M, Kaczorek D, Kuzyakov Y, Breuer J. Silicon Pools and Fluxes in Soils and Landscapes-A Review. J Plant Nutr. Soil Sci. 2006; 169:310-329.

56. Subramannian S, Gopalswamy A. Influence of silicate and phosphate materials on availability and uptake of silicon and phosphorus in acid soils. Oryza. 1990; 27(3):267-273.

57. Talashilkar SC, Chavan AS. Effects of rice hull ash on yield and uptake of silicon and phosphorus by rice cultivars at different growth stages. Journal of the Indian Society of Soil Science. 1996; 44(2):340-342.

58. Thiagalingam KJ, Silva A, Fox RL. Effect of calcium silicate on yield and nutrient uptake in plant growth on a humic ferruginous latosol. In: Proc. Conf. On chemistry and fertility of tropical soils. Kualalumpar, Malaysia. Malaysian Society of Soil Science, 1977, 149-155.

59. Tisdale SL, Nelson WL, Beatson JD. Soil fertility and fertilizers. MacMillan Publ. New York, 1985, 754.

60. Vandervorm PDJ. Uptake of Si by five plant species as influenced by variations in Si-supply. Plant and Soil. 1980; 56:153-156.

61. Vyas MK, Motiramani DP. Effect of organic matter, silicates and moisture levels on availability of phosphate. Journal of the Indian Society of Soil Science. 1971; 19(1):39-43.

62. Walker. Calcium silicate enhances resistance of sugarcane to the African stalk borer *Eldana saccharina* (Lepidoptera: Pyralidae). Agricultural and Forest Entomology. 2002; 4(4):265-274.

63. Webley DM, Henderson MEK, Taylor IF. The microbiology of rocks and weathered stones. J Soil. Sci. 1963; 14:102-112.

64. Webley DM, Duff RB, Mitchell WA. A plate method for studying the breakdown of synthetic and natural silicates by soil bacteria. Nature, 1960; 188:766-767.

65. Yoshida S. The physiology of silicon in rice. Food fertilizer tech. Centre Technical Bull, 1975, 25.

Chapter - 7
Host Plant Resistance in Legumes against Bruchids Sp.

Authors

Soumia P.S

[1] Division of Crop Protection, ICAR-Directorate of Onion and garlic Research, Pune, Maharashtra, India

Chitra Srivastava

Division of Entomology, Indian Agricultural Research Institute, New Delhi, India Instead of Directorate of Onion and Garlic Research, India

Subramanian S.

Division of Entomology, Indian Agricultural Research Institute, New Delhi, India

Guru Prasanna Pandi G.

Division of Crop Protection, ICAR-National Rice Research Institute, Cuttack, Odisha, India

Chapter - 7

Host Plant Resistance in Legumes against Bruchids Sp.

Soumia P.S, Chitra Srivastava, Subramaniam S and Guru Prasanna Pandi G.

1. Introduction

Grain legumes/pulses are the second most important group of crops after cereals, particularly in developing countries where the availability and consumption of animal protein is rather low due to various socio-economic constraints. Pulses are excellent and cheapest source of dietary proteins (25-40%), carbohydrates (50-60%), fat, minerals, vitamins and certain amino acids like lysine and tryptophan (Lambrides and Godwin, 2007). This makes them a good supplement to the cereal based diets; which are usually rich in carbohydrates but low in proteins. According to recommendation of the World Health Organization, a person must consume about 80 g of pulses per day in India. Pulses are grown on 68.3 million hectares and furnish 57.5 million tonnes to the world's food basket. The major pulse crops grown in India are chickpea, pigeonpea, mung bean, urdbean, cowpea, lathyrus, horsegram, lentil and peas. India is the world's largest producer of pulses with a total production of about 14.7 million tonnes from about 23.4 million hectares, but the average productivity (550-625 Kg/ha) is quite low (Economic Survey, 2011). Unless the country witnesses a major technological advance in pulse cultivation to meet the domestic demand, imports of pulses seem inevitable. The slow growth in pulse production compared to burgeoning population had led to progressive decline in per capita availability of pulses from 60 gm/day/person in 1951 to 31 gm/day/person in 2008.Legumes are severely affected by storage insect pests predominantly in the Tropics and Sub-tropics where temperatures and relative humidity are high that cause both qualitative and quantitative loss. Unlike field pests, stored grain pests are more serious not only due to the magnitude of loss they cause but also by the irreversibility of the damage they cause (Keneni *et al.*, 2011).

1.1 Role of Pulses in Food Security and Associated Constraints in Pulse Production

Pulses containing 25 per cent protein, which is approximately thrice to that of cereals, play a fundamental role in the food security of most developing countries. Pulses are also a low-fat, fibre rich source of protein and an indispensable part of traditional food baskets. It is included in all dietary guidelines especially in the World Food Programme (WFP) to eliminate 'protein calorie' malnutrition (Kakati *et al.,* 2010). Hence, a significant part of human population relies on pulses that can be complemented with cereals. Moreover, of the total pulse production, about 60% is deployed mainly for human consumption with more utilization in low income countries.

On the other hand, the world has entered an era of "peak food" production, with almost stagnant production of 11 and 14 million tonnes over the last two decades. One third of world's population that resides in India will face serious food shortage as the burgeoning population would reach 1.68 billion by 2030. Although India is the world's largest producer of pulses, yet 3.5 million tonnes of pulses were imported during 2009-10. Consequently, it has been projected that India may require 32 million tonnes by 2030 to meet its domestic demand (http://www.icar.org.in/files/ICAR-Vision-2030.pdf).

Most pulses producing countries are still under developed. It is very difficult to transfer new technologies to these areas, which become a major constraint on pulses production. Leading pulse producing countries such as Bangladesh, India, Iran, Myanmar, Malaysia, Mexico, Turkey and Syria also face serious abiotic (soil fertility, drought, hurricanes, flood incidence, and unpredictable weather) and biotic problems (pests and diseases). Besides, it is evident that in near future, pulse cultivation in these areas would face more problems thereby leading to inconsistency in production, import and export. Even with the best efforts, pulses production is stagnant as pulses are grown mainly as residual crops on marginal lands. Furthermore, under changing climatic condition yield loss by diseases and pests outbreaks in pulses will increases (Srivastava, 2002).Among the various biotic problem that invade pulse production, pests and diseases alone might be responsible for multi-million rupees economic losses.It has been predicted that 2-2.4 million tonnes of pulses are ravaged by pests with monetary loss of Rs. 6,000 crores annually (Reddy, 2010).In India, over 200 insect pests have been reported to attack various pulses both in the field and storage (CABI, 2007). However, the present study exclusively focuses on *Callosobruchus* spp. which is responsible for serious post- harvest loss in pulses in during storage.

1.2 Post Harvest Losses in Pulses

Post-harvest losses have posed a serious question to food security as it causes 9.5% loss in total pulse production. Among the different losses in post-harvest operation, storage holds a major share (7.5%) followed by processing (1%), transport (0.5%) and threshing (0.5%). Deshpande and Singh (2001) reported that pulses (5%) were most vulnerable to storage insect pests in comparison to cereals like maize (3.5%), wheat (2.5%) and paddy (2%). Insect pests readily attack pulses due to the presence of high protein content. Controlling these post-harvest losses is obligatory to reduce the import quantity of our nation. According to FAO's prediction approximately 70% of the farm products are stockpiled by farmers for various purposes like their own consumption, animal feed, seed and other purposes. Under Indian situation, farmers generally use locally designed storage structures for bulk store of their commodities which in most cases are not insect or moisture proof (Das *et al.,* 2013). Pulse beetle, being cosmopolitan can thrive well in a variety of pulses and hence it is very difficult to control this obnoxious pest. It has been documented that pulse beetle can cause up to 30% loss within a short period of time (Raja and John, 2008). They inflict both qualitative and quantitative loss to legumes, making them unfit for sowing (loss viability) or for food or feed (due to spoilage and bad smell) (Aslam *et al.,* 2006; Haile, 2006).

1.3 Pulse Beetle – *Callosobruchus* spp.

Callosobruchus spp.is the most important field carry over storage pest of pulses such aschickpeas, lentils, green gram, broad beans, soybean (Desroches *et al.,* 1995; Bushara, 1988; Talekar, 1988; Kashiwaba, *et al.,* 2003) adzuki bean and cowpeas (Udayagiri and Wadhi 1989). Primary infestation starts in the field, where adult beetles deposit their eggs on matured pods which are later carried over to the storage where substantial losses may occur within three months (Singh and Jackai, 1985). Damage made by this pest mainly due to the feeding of the larvae inside the seed, which is important because of overall weight loss; altered nutritional, aesthetic quality and germination potency of the seed (Talekar, 1988). Bruchids in stored seed are a major problem because of their ability to re-infest stored seed. A single beetle is able to cause 3.5% weight loss in cowpea seeds (Brooker, 1967). *Callosobruchus* spp. originated in Africa and is distributed throughout the tropics and sub-tropics region of Asian and African countries (http://www.cabi.org). The spread of this species occurred with the movement of harvested legumes, either from the paddock into storage, or in contaminated seed within and between storage facilities (http://www.padil.gov.au/pbt).

There are number of *Callosobruchusspp.*found to be a pest for pulses. *C. maculatus, C. analis* and *C.chinensis*are the most dominant pest species to the pulses found in tropical and sub-tropical countries of Asia and Africa. Male and female beetles of *Callosobruchus* spp. are easily distinguished by their general appearance. Generally, females are stout and larger than males, but there is much variation.The most unique characteristic is the cuticular colour of the abdominal plate (pygidium) covering at the rear end. In some strains, females are black in colorations and males are brown, but in others both sexes are brown (Southgate, 1958).

In general *Callosobruchus* spp has an optimum developmental conditions about 23-35°C and 70 per cent RH, but may vary depending upon species and weather parameters (http://www.cabi.org). Of the suitable host and optimum development conditions (30° C and 70% RH), the minimum development period for *C. maculatus* is about 24 days and 22.3 days for *C. chinensis* (Raina, 1970).

2. Different Tactics to Manage Pulse Beetle

In general, farmers are forced to sell their produce shortly after harvest due to inevitable losses. Therefore, there is a need to develop sustainable ways to manage this belligerent pest. Several options are available for controlling stored grain pests. Control can be at two levels i.e., domestic / small scale and large commercial level. Although various control measures (chemicals, dusts, oils, fumigants, irradiation etc.) are available, it is often not ideal to add any foreign substance to the produce in close proximity to the time of consumption. This may invite problems related to persistent toxic residues, insect resistance and adverse environmental impact.

2.1 Host Plant Resistance

Host plant resistance, where bruchid resistant plants would be a better option and can be integrated with other control measures, provided a promising impact on bruchid management in stored pluses. Thus, development of resistant cultivars to bruchids is a major target in mung bean breeding projects. Fujii and Miyazaki (1987) identified a wild mung bean (*V. radiata* var. *sublobata* (Roxb.) Verdc.). That had been exhibiting complete resistance to both *C. chinensis* and *C. maculates* and was widely used inmung bean breeding programme (Kitamura *et al.,* 1988).

During the course of evolution, both insects and plants continuously interact with each other in a complex way. During the mutual relationship insects provide pollination and defense whereas, plants offer food, shelter and oviposition sites for insects (Panda and Khush, 1995). Conversely, insect pest might be harmful to plants based on the intensity of attack leading to their death. Plants overcome insect attack through various defense mechanisms *viz.*, physical and biochemical barriers such as trichome density, waxy coating on the leaves, seed coat hardness, stimulation of defensive proteins, volatiles and secondary metabolites (Fordyce and Agrawal, 2001; Birkett *et al.,* 2000; Kliebenstein *et al.,* 2001). This phenomenon is commonly called as co-evolution or reciprocal selection.

Painter (1951) described host plant resistance (HPR) as "those characters that enable a plant to avoid, tolerate or recover from attacks of insects under conditions that would cause greater injury to other plants of the same species". HPR is eco-friendly as well as compatible with other control methods and might synergise their effects (Eigenbrode and Trumble, 1994). The innate protection of resistant plants disrupts the normal association of the pest with its host plant (Dicke, 2000). The processes of resistance involve morphological, physiological and/or biochemical mechanisms which range from simply minimizing the effect of insect attack to adversely affecting the insects cellular processes, growth and development (Gatehouse, 2002). Varietal resistance to control storage pests could provide a base for developing an integrated pest management (IPM) programme (Semple, 1992; Dobie, 1984). Edwards and Singh (2006) proposed that legumes employ a range of direct (morphological barriers) and indirect (secondary metabolites and anti-nutritional compounds) protection. Seed of a resistant host plant produce adverse effects on stored pest larvae during feeding may also have combination of morphological and biochemical features of the host plant (Panda and Khush, 1995).

2.2 Mechanism of Resistance against Pulse Beetles

The ecological relationship between plants and insects is a complex one with physical as well as chemical interactions. It is argued that plant resistance is not dependent on a single mechanism, but there are overlaps between the morphological and biochemical bases of resistance. The relationship between bruchids and their host plant is complex; therefore, the adaptation mechanism to these pests is difficult (Taylor, 1981). In this situation, through knowledge of the underlying mechanisms of bruchid resistance is vital for developing suitable breeding strategies. As far as the mechanism of seed resistance against bruchids is concerned, many strategies are used by seeds to protect themselves against insects: i) seed may be too hard for newly hatched larvae to penetrate; ii) seed may be physically too small or with an inconvenient shape for larvae to reach full size; iii) seed may contain too little food to support the larvae; iv) seed may contain toxins or other substances in the cotyledon or its enveloping seed coat that inhibit the larval development (Kashiwaba *et al.,* 2003).

Paniter (1951) proposed three mechanisms of HPR *viz.*, antixenosis (non-preference), antibiosis and tolerance which are worldwide accepted. In general, Non-preference type is refractory to the pest thereby avoiding the insect from using the plants for feeding, egg laying (oviposition) and shelter or combination of these three processes. Non preference type of resistance has been reported in maize varieties against weevils (Kang *et al.,* 1995; Derera *et al.,* 2001). Antibiosis is differed from antixenosis by allowing the insect to colonize the host plants, but provide resistance after ingestion of host tissues by producing adverse effect on insect's growth, reproduction and survival (Dent, 2000). Phytotoxins (primary metabolites) and secondary metabolites are associated with antibiosis. Earlier workers have reported that phenolics in grains (hydoxycinnamic acids) act as an important resistance factor to storage pests (Classen *et al.,* 1990; Arnasson *et al.,* 1992; Sen *et al.,* 1994) which clearly indicated that biochemical analysis needs to be considered for screening bruchid resistance genotypes. Tolerance is the ability of the host to rapidly recover, repair or withstand infestation. However, tolerance is not applicable in storage condition because the damage caused to stored grains is irreversible (Horber, 1989). Therefore, antixenosis and antibiosis are highly important in this study.

2.3 Bruchid Resistance Testing Methods

McCain *et al.,* (1964) claimed that the quantity of damage is to be taken as a measure of resistance. Therefore, scientists have exploited different screening methods to test resistance. Under storage condition, basically two methods are used to check resistance against stored pests that are free-choice and no-choice (force) tests. In free-choice test, the insects can infest the seed samples of their choice (Giga, 1995). The ability of a specific variety to repel insects (antixenosis) is usually measured with this method. This method is mostly used when a greater number of germplasm to be screened. Horber and Mills (1976) suggested that free-choice test method is suitable to eliminate most susceptible genotypes. Next method, no-choice test, where the insects are restricted in their choice of seed sample (Nwanze and Horber, 1975; Giga, 1995). This method is widely used in laboratory screening of genotypes for resistance against storage pests. Here, antibiosis plays a major role in resistance, which is measured in terms of the percent weight loss, mean development period and percent insect emergence. Further, all these parameters are used for calculating growth index. Both of these methods have been previously used to screen germplasms against bruchid resistance (Tipping *et al.,* 1989; Cardona *et al.,* 1990; Derera *et al.,* 2001; Edde and Amatobi, 2003; Cardona *et al.,* 2005).

3. Screening of Different Pulses against Pulse Beetles

Many varieties of cowpea screened for resistance against the cowpea bruchid, *C. maculatus* and other *Callosobruchus*spp. Showed that some varieties possess a very high level of resistance (Jackai and Daoust, 1986). Ahmed *et al.* (1993) reported that coefficients of phenotypic and genotypic variations were positively correlated with damaged seeds and emergence holes.Ashraf *et al.* (1985) tested the cultivars of chickpea (86208, CM-72, 86221, 86037 and C-44) for their relative resistance against this beetle and found that C-44 was relatively susceptible whereas 86037 appeared to be resistant against the beetle. Edward and Gunathilagaraj, (1994) screened 26 accessions of chickpea for their resistance against pulse beetle and concluded that resistance was due to antibiosis as reflected in lower survival, prolonged development period and adults with reduced longevity.

Singh and Sharma (2001) evaluated thirteen varieties of chickpea for oviposition preference and subsequent larval development of the pulse beetle *C. chinensis*. PG-5 was the most resistant variety with minimum GI of 1.358 and longest grub development period of 28.33 days. On the other hand, GNG-663 was found to be the most susceptible with maximum GI of 2.211 and minimum grub development period of 28 days.Singh and Sharma (2003) evaluated seven varieties *viz.*, L-4076, DPL-61, K-75, K-303, Sehore-747, DPL-59 and L-414-7 of lentil. The pulse beetle showed high ovipositional preference for all the varieties, as grain with eggs in the varieties ranged from 97.0 to 99.6 per cent. Of the varieties evaluated, DEP-3 was observed to be the most preferred one with GI of 1.376; the grub developmental period in the variety was 33.66 days. The variety RPG-3 had minimum GI of 0.671 (developmental period, 38.00 days). The pest GI ranged from 1.051 (BR-12) to 0.805 (L-116) in the other varieties.

Kananji (2007) evaluated forty-two genotypes of Malawian dry bean germplasm for resistance to *Z. subfasciatus*. Significant differences were observed in grain weight loss (%), number of adult bruchids emerged and Dobie susceptibility index among the genotypes. The relationship between grain weight loss (per cent) and adult bruchid emergence was highly significant ($P<0.001$) and positively correlated ($r = 0.803$). There was also a highly significant ($P<0.001$) and positive correlation ($r = 0.734$) between weight loss (per cent) and Dobie's susceptibility index (DSI).Obiadalla-Ali *et al.* (2007) screened 21 cultivars of cowpea for resistance to weevil pest based on pest development assessment including number of adults emerged, percentage of damaged seeds, developmental period (days), number of eggs laid on seeds, mean of eggs laid per female, average lifespan of female (days), per cent adult emergence and per cent loss in seed weight. The assessed cowpea cultivars were classified into three groups, sensitive (Dokki 331 and Creamy 7), moderately tolerant (Black eye and IT 90 K 2840-2), highly tolerant (IT 93 K 12904 and IT 81 D-1064).Laserna-Ruiz *et al.* (2012) screened a lentil germplasm collection of 571 accessions from 27 different countries including wild species for susceptibility to seed bruchids under natural field conditions in Central Spain. A total of 32 accessions, including *L. culinaris* subsp. *culinaris*, *L.culinaris* subsp. *orientalis, L. nigricans*, and *L. lamottei* showed lower infestation rates than the check and were selected as potential sources of resistance to this pest.

Relative resistance study of 12 chickpea genotypes against of *C. maculatus* was carried out. The genotypes CH-52/02 and B-8/03 were found to be least susceptible to bruchids attack. The tolerant genotypes were dark brown coloured, hard seeds with wrinkled seed coat and small sized grains (Sarwar, 2012).Sarwar (2013) studied the relative susceptibility among eight lentil genotypes against *C. analis.* Feeding preference, seed weight loss, frass weight, reproduction and developmental period of the pest, seed coat thickness and germination percentage was estimated to determine the differential response of lentil genotypes. Genotype AEL-49/20 displayed relatively tolerant response than AEL-23/40.

4. Resistance Factors to Storage Pests

The present knowledge about crop resistance factors against stored pests had mainly focused to establish the factors which confer resistance. Understanding the mechanism will help plant breeders to effectively incorporate these resistance factors as an integral part along with other desirable traits for developing new varieties which inherently express resistance against storage pests. In general, for storage pests, only physical and biochemical factors are more influential that govern resistance (Semple, 1992).

4.1 Physical Factors

The physical factor like seed coat roughness and seed coat hardness confer resistance to pulse beetle (Van Huis and De Rooy, 1998). Seed with hard coat may prevent the larvae from effectively piercing the seed, while a rough seed coat makes oviposition difficult (Nwanze and Horber, 1976; Messina and Renwick, 1985). Similar trend of results obtained in cowpea against *C. maculatus* (Shade *et al.*, 1996) in chickpea and brad bean against *C. chinensis* (Brewer and Horber, 1983).Talekar and Lin (1992) investigated characteristics of resistance to C. chinensis in two mung bean and one black gram accessions and concluded that smaller seed size of the accessions was not responsible for the resistance. The results indicated the presence of anti-nutritional factors in the resistance accessions. Cultivars with hard seed coat showed nonpreference by pulse beetle (Ahmed *et al.*, 1993).

Shade *et al.* (1996) reported that seed with tough seed coat may hinder the larval penetration into the seeds to an extent, though primary feeders are well adapted to chew through.Twenty-six mung bean (*V. radiata*) varieties and accessions were screened for resistance to four bruchid species (Coleoptera: Chrysomelidae). On the basis of percentage of seeds damaged, commercial mung bean varieties tested were highly susceptible to *C. maculates* and *C. chinensis*. Three accessions (TC1966, ACC23 and ACC41) showed bruchid resistance due to small seed size and the presence of a well-formed texture layer on the seed. The results revealed that small seed size in mung bean may act as an oviposition deterrent to bruchids (Lambrides *et al.*, 2000).

The presence of anti-nutritional factors in the cotyledons of resistant accessions of *V. umbellata* confirmed that the bruchid larvae were able to penetrate through the seed coat into cotyledons but, failed to support a normal adult emergence. This was due to the mortality of first and second instars larva inside cotyledons of the seed. The result suggested that the seed coat of all resistant accessions had no protective role against the bruchids (Kashiwaba *et al.*, 2003). Landerito, *et al.* (1993) studied the resistant and susceptible accession of mung bean by seed hardness. The resistant seed varieties were found to be harder than susceptible variety.

4.2 Biochemical Factors

In many legumes, the chemical defense system is mediated mainly by several chemicals of one class or compounds of several classes to increase the defense levels. This indicates that a number of specific insect resistance mechanisms have been identified in these plant groups although they have not yet been widely integrated into mainstream breeding programs (Edwards and Singh, 2006). High resistance of some of the cultivated or wild *Vigna* spp. to *C. maculatus* may be due to the presence of multiple chemical factors with additive action to protect seeds from predation. Trypsin inhibitors

(Savelkoul *et al.,* 1992) and tannins present in the seed coat (Deshpande, 1992) of legumes confer resistance against bruchids. Lattanzio *et al.,* (2005) reported that the resistance against C. *maculatus* in cowpea is contributed by condensed tannins (Proanthocyanidins). Lale and Kolo (1998) documented that certain biochemical factors in the testa/seed coat of resistant cowpea genotypes inhibit the oviposition and survival of pulse beetle eggs irrespective of seed coat texture. Biochemical factors in the seed coat might influence the egg hatching (Lale, and Makoshi, 2000). However trypsin inhibitor, tannin and starch content present in the seeds found to have significant positive correlation with bruchid resistance (Marconi, 1997). Ali and Smith (2000) reported that in the decorticated faba seeds allowed for the higher percentage of larval penetration and subsequent growth and emergence than the whole seeds. It is indicating that a characteristic of seed coat and biochemical antibiosis provides resistance in faba bean varieties. However, Edde and Amatobi (2003) observed that bruchids preferred whole cowpea seeds for oviposition compared to decorticated seeds. Therefore, in the cowpea seeds, testa/seed coat may not be a valuable aspect to confer bruchid resistance.

Cardona *et al.* (1989) further confirmed that the seed coat may act as physical barrier; the actual factors that are responsible for bruchid resistance were present in the cotyledon and are chemical in nature. Thus, it is evident that bruchid resistance under storage condition is due to the interaction of antibiosis and non-preference. *Phaseolus vulgaris* differ in their resistance reaction to *A. obtectus* and *C. chinensis* due to differential absorption of soluble heteropolysaccharides like arabinose, galactose, glucose, raffinose and xylose (Suzuki *et al.,* 1995). Bhattacharya and Banerjee (2001) reported that the penetration capacity of the neonate larvae of bruchid beetle was inversely related to the phenol contents of the seed. According to Horber (1983), breeding with genotypes containing more concentrations of heteropolysacharide may confer resistance against bruchid beetles. Other biochemical factors present in the seed like tannins (seed coat), arcelin and phytohemagglutinin (PHA) (cotyledons) including α-amylase inhibitors also conferred resistance in dry beans (Schoonhoven *et al.,* 1983; Osborn *et al.,* 1986; Ishimoto and Kitamura, 1989; Gatehouse, 1990). All of these studies demonstrated the importance of antibiosis as main source of resistance that affects the growth and development of larvae (Gatehouse *et al.,* 1979).

Susceptible and resistant accessions of mung bean and black gram were screened against bruchid resistance. Factors like crude fibre, silica and lignin contents were same in both resistant and susceptible varieties. Saponins were not detected while alkaloids were present in both the susceptible and resistant accessions. Bruchid resistant varieties had higher trypsin inhibitor activities than susceptible varieties. Moreover, polyphenol contents in the seeds were not correlated with bruchid resistance. Albumin and globulin fractions of protein adversely affected egg deposition of bruchid beetles (Landerito, 1993). Cowpea weevil found difficulty to developing even with the low concentration (1%) of bean α amylase inhibitor (Pedra *et al.*, 2003). Similarly, Ishimota and Kitamura (1989) and Shade *et al.* (1994) reported the detrimental effect of α amylase inhibitor in the growth and development of *C. maculatus* and *C. chinensis*.

There is sufficient body of literature explaining the bean resistance to mechanism of bruchids resistance under controlled laboratory conditions (Hartweeck *et al.*, 1991; Fory *et al.*, 1996). The presence of the arcelin, seed hardness, seed coat thickness, tannins, lectins and trypsin inhibitors have been used to explain antibiosis or no preference resistance mechanisms to bruchids infestation. Bhattacharya and Bancrjee (2001) studied the effects of physical parameters in large garden pea, chickpea, and the texture and thickness of the host preference, shape, colour and size of pulse seeds and the texture and thickness of the seed coat affected egg laying. Moisture contents in lentil seeds differed significantly ($P<0.05$) that affected egg laying and fecundity of the female.

5. Heritability of the Physical Parameters

Fifty six mung bean accessionswere evaluated to determine the variability, heritability between the agronomic and physiological characters associated with yield. Heritability was found to be high for days to flowering and pod length (Khajudparn, and Tantasawat, 2013). Saeed *et al.* (2007) studied for the association of seed yield with some important morphological characters in forty three mung bean accessions. High heritability was observed for seed yield, node of the first peduncle, pod clusters on branches as well as per plant.Rahim *et al.* (2010) evaluatedheritability, variability and genetic advance in26 mung bean genotypes. High broad sense heritability and genetic advance was observed for some plant traits like plant height, number of pod, number of seeds per pod, 1000-seed weight and grain yield. Twenty mung bean genotypes were evaluated for genetic variability and correlation among different yield contributing traits. Heritability was high for pod length and plant height, while pods per plant and seed per pod had low heritability (Zaid *et al.*, 2012).

Pure line of mung bean seeds was treated with four doses of ethyl methane sulphonate to analyze the genetic advance and heritability of some major agro-metrical traits in second generation mutant. PCV was found to be higher than GCV for all the traits studied. Low heritability (38.43) and genetic advance (5.37) were noticed in control for the total seed yield per plant (Roychowdhury *et al.,* 2012). Eighty-eight finger millet germplasms were tested to find out the heritability, variance components and genetic advance for some yield related agronomic traits. The variance components like the phenotypic and genotypic variance were found to be high. Heritability coupled with expected genetic advance was recorded high for number of tillers per plant, number of ears per plant, number of finger per plant, finger length, grain yield and thousand seed weight (Wolie *et al.,* 2013).Genetic variability was studied in forty mung bean genotypes by Narasimhulu *et al.* (2013).High GCV and PCV were found for number of branches per plant, pods per plant, and yield per plant and harvest index. High estimates of genetic advance as percent of mean were observed for 100 seed weight and harvest index.

6. Artificial Seed System

An artificial seed system was used to evaluate the effect of allelochemicals and other plant fractions on growth and development of bruchid, *C. maculatus*. Artificial seeds were prepared by using different proportions of various inhibitors of mammalian digestive proteases with the flour of susceptible cowpea cultivar. This experiment demonstrated the applicability of the technique to evaluate the effect of different chemicals on bruchid growth and development (Shade *et al.,* 1986). Evaluation of resistance using artificial seeds showed that increasing the percentage of resistant seed powder adversely affected bruchid growth and development. Both resistant groups varied in resistance when compared to susceptible cultivar. The results suggest that biochemical (s) in cotyledon tissue are responsible for the resistance and the seed coat had no protective role against the bruchids (Somta *et al.,* 2008). When artificial seeds made of *V. umbellate* flour were used, decreased adult emergence and prolonged mean developmental period of*C.maculatus* was observed. The biochemical constituents present in the cotyledon tissue had a negative correlation with that of index of suitability. The anti-nutritional factors present in the cotyledon tissues affected the adult emergence in *C.maculatus* (Kashiwaba *et al.,* 2003). A polysaccharide from seeds of common bean, *P. vulgaris*, was found to be toxic to *C. chinensis* as evident by increasing larval mortality, reduced growth and development and fewer eggs (Applebaum *et al.,* 1970). More or less same number of eggs was laid on the artificial seeds (control) in comparison to black-eyed beans, yet per cent survival from artificial seeds was abridged. Hardness of the artificial seeds prejudiced oviposition behavior of the beetle with more eggs laid on the harder seeds (Hudaib *et al.*, 2013).

7. Genetic Diversity of *Vigna* spp

Characterization and assessment of qualitative and quantitative traits among *Vigna* spp. have great significance in designing breeding strategies (Li *et al.*, 2001). Usually the morphological traits, pedigrees, geographic origins, isozymes and DNA markers are used for the assessments of the genetic diversity *Vigna* spp. (Fatokun *et al.*, 1992; Kaga and Ishimoto, 1998; Bisht *et al.*, 2005). However, the morphological characters are limited in number and are often subjected to constraints from environmental and epistatic interactions (Tantasawat, *et al.*, 2010). To overcome these environmental discrepancy, the genetic markers such as restriction fragment length polymorphisms (RFLPs), random amplified polymorphic DNAs (RAPDs), amplified fragment length polymorphisms (AFLPs), microsatellites or simple sequence repeats (SSRs) and inter-SSR (ISSR) are used to the characterization of genotypes in *Vigna* spp. (Tantasawat, *et al.*, 2010; Sarwat, 2012). In addition, they also allow variety identification in the early stages of development (Tan *et al.*, 2012). RAPD, AFLP, SSR and ISSR used widely for estimation of genetic variability in *Vigna* spp. (KagaandIshimoto, 1998; Lakhanpaul *et al.*, 2000; Chattopadhyay *et al.*, 2005). The SSR markers consist of randomly repeated units of short nucleotide motifs with 1 to 6 bp long. Although SSRs are highly polymorphic and their single-locus nature still limit their uses for genetic diversity study in several plant species. But several reports had been found the SSR markers were successfully yielding the reproducible pattern of genomic fragments in cowpea (Li *et al.*, 2001), mung bean (Kumar, 1999; Tan *et al.*, 2012) and azuki bean (Wang *et al.*, 2004).

A comparative study of marker systems in *Vigna* spp. showed that SSR marker was more efficient in detecting genetic variability when comparing with other marker in mung bean and cowpea (Dikshit *et al.*, 2007). Bhuyan *et al.* (2014) analyzed the diversity in Indian mung bean germplasm with SSR, RAPD, and ISSR markers and the result suggested that ISSR and RAPD were quite effective in resolving the genetic variability in the studied genotypes.As well the single locus (SSRs) and multilocus analysis (RAPD and ISSR) of DNA was not only useful for genotyping, but could also be used for mapping, gene tagging and marker-assisted selection of the superior traits. Genetic linkage maps of mung bean have been constructed using RFLP, RAPD and SSR markers (Fatokun *et al.*, 1992; Menancio-Hautea *et al.*, 1993; Lambrides *et al.*, 2000; Humphry *et al.*, 2002;Humphry *et al.*, 2003; Humphry *et al.*, 2005; Chankaew *et al.*, 2011; Young *et al.*, 1993; Kagaand Ishimoto, 1998; Mei *et al.*, 2009) have been used to map genes for azuki bean weevil resistance (Young *et al.*, 1993; Kaga and Ishimoto, 1998), seed color (Lambrides *et al.*, 2004), to identify QTLs for seed weight (Fatokun *et al.*, 1992; Humphry *et al.*, 2005; Mei *et al.*, 2009) and hardness of seed (Fatokun *et al.*, 1992).

8. Conclusion

To date, entomological studies conducted in India have concentrated more on field than storage pests and previous breeding programmes have not focused on developing bruchid resistant lines. This has resulted in the development of superior varieties with high levels of resistance or tolerance of other biotic and abiotic constraints, but is more vulnerable to bruchid invasion in storage. Genetic diversity is an important factor for crop improvement programme and its nature and degree of variability would be useful for selecting desirable parents from available germplasm for a successful breeding programme. Despite the fact that many workers have conducted varietal screening for bruchid resistance, there is further need to identify resistant varieties, as bruchids have inherent ability to tackle the resistance mechanisms.

References

1. Ahmed K, Khalique F, Khan IA, Afzal M, Malik BA. Genetic differences for susceptibility of chickpea to bruchid beetle (*Callosobruchus chinensis* L.) attack. *Pakistan Journal* of Scientific and *Industrial* Research. 1993; 36(2-3):96-98.

2. Ali K, Smith RH. The effect of seed coat on the susceptibility of faba bean to *Callosobruchus chinensis* L. African Crop Science Journal. 2000; 9:421-430.

3. Applebaum SW, Tadmor U, Podoler H. The effect of starch and of a heteropolysaccharide fraction from *Phaseolus vulgaris* on development and fecundity of Callosobruchus maculatus. Entomologia Experimentalis ET Applicata. 1970; 13:61-70.

4. Arnason JT, Gale J, de Beyssac BC, Sen A, Miller SS, Philogene BJ *et al.* Role of phenolics in resistance of maize grain to the stored grain insects, Prostephanus truncatus (Horn) and Sitophilus zeamais (Motsch.). Journal of Stored Products Research. 1992; 28(2):119-126.

5. Ashraf M. Farming systems approach: Research on cowpeas and

extension. In: eds. Singh SR. and Rachie KO. Cowpea Research, Production and Utilization. Wiley, New York, 1985, 341-349.

6. Aslam M, Shaheen FA, Abbas MA, Saba A. Management of *Callosobruchus chinensis* Linnaeus through use of resistance in stored chickpea varieties. World Journal of Agricultural sciences. 2006; 2:82-84.

7. Bhattacharya B, Banerjee TC. Factors affecting egg-laying behavior and fecundity of *Callosobruchus chinensis* (L.) (Coleoptera: Bruchidae) infesting stored pulses. Oriental Insects. 2001; 35(1):373-386.

8. Bhuyan SI, Hossain MS, Islam MM, Begum SN, Urbi Z, Hossain MS. Molecular Assessment of Genetic Diversity and Relationship in Selected Mungbean Germplasms. Biotechnology. 2014; 13(3):126.

9. Birkett MA, Campbell CAM, Chamberlain K, Guerrieri E, Hick AJ. New roles for cis-jasmone as an insect semiochemical and in plant defense. Proceeding of national academy of science. 2000; 97:9329-9334.

10. Bisht IS, Bhat KV, Lakhanpaul S, Latha M, Jayan PK, Biswas BK *et al.* Diversity and genetic resources of wild Vigna species in India. Genetic Resources and Crop Evolution. 2000; 52(1):53-68.

11. Booker RH. Observations on three bruchids associated with cowpea in northern Nigeria. Journal of Stored Products Research. 1967; 3:1-15.

12. Brewer INM, Horber E. Evaluating resistance to *Callobruchus chinensis*. Linn. In different legumes. In Proceedings of Third International Conference on Stored Product Entomology, October, Manhattan, Kansas, USA. 1983, 23-28, 435-443.

13. Bushara AG. Insect depredation during Storage. In World Crops: Cool Season Food Legumes, the Netherlands, 1988, 367-378.

14. CABI. Crop Protection Compendium. Wallingford, UK, CAB International, 2007.

15. Cardona C, Kornegay J, Posso CE, Morales F, Ramirez H. Comparative value of four arcelin variants in the development of dry bean lines resistant to the Mexican bean weevil. Entomologia Experimentalis ET Applicata. 1990; 56:197-206.

16. Cardona C, Posso CE, Kornegay J, Valor J, Serrano M. Antibiosis effects of wild dry bean accessions on the Mexican bean weevil and the bean weevil (Coleoptera: Bruchidae). Journal of Economic Entomology.

1989; 82(1):310-315.

17. Cardona C, Valor JF, Bueno JM, Mejia A, Blair M. Levels of resistance to important insect pests confirmed in bean progenies. Highlights of CIAT Research Activities Annual report, CIAT, Cali, Colombia, 2005.

18. Chankaew S, Somta P, Sorajjapinun W, Srinivas P. Quantitative trait loci mapping of *Cercospora* leaf spot resistance in mung bean, Vigna radiata (L) Wilczek. Molecular Breeding. 2011; 28:255-264.

19. Chattopadhyay K, Ali MN, Sarkar HK, Mandal N, Bhattacharyya S. Diversity analysis by RAPD and ISSR markers among the selected mung bean [*Vigna radiata* (L.) Wilczek] genotypes. The Indian Journal of Genetics and Plant Breeding. 2005; 65(3):173-175.

20. Classen D, Arnason JT, Serratos JA, Lambert JDH, Nozzolillo C, Philogene BJR. Correlation of phenolic acid content of maize to resistance to *Sitophilus zeamais*, the maize weevil, in CIMMYT'S collections. Journal of Chemical Ecology. 1990; 16(2):301-315.

21. Das I, Kumar G, Shah NG. Microwave heating as an alternative quarantine method for disinfestation of stored food grains. International Journal of Food Science, 2013. doi:10.1155/2013/926468.

22. Dent JA, Smith MM, Vassilatis DK, Avery L. The genetics of ivermectin resistance in Caenorhabditis elegans. Proceedings of the National Academy of Sciences. 2000; 97(6):2674-2679.

23. Derera J, Pixley KV, Denash G. Resistance of maize to the maize weevil: I. Antibiosis. African Crop Science Journal, 2001; 9(2):431-440.

24. Deshpande SD, Singh G. Long Term Storage Structures in Pulses, National Symposium on Pulses for Sustainable Agriculture and Nutritional Security. Indian Institute of Pulses Research, Kanpur, 2001, 17-19.

25. Deshpande SS. Food legumes in human nutrition: A personal perspective. Critical Review on Food Science. 1992; 32:333-363.

26. Desroches P, Elshazly E, Mandon N, Duc G, Huignard J. Development of *Callosobruchus chinensis* (L.) and C. *maculatus* (F.) (Coleoptera: Bruchidae) L. in seeds of *Vicia faba* differing in tannin, convicine and vicine contents. The Journal of Stored Products Research. 1995; 31:83-89.

27. Dicke M. Chemical ecology of host-plant selection by herbivorous arthropods: a multitrophic perspective. Biochemical Systematics and

Ecology. 2000; 28(7):601-617.

28. Dikshit HK, Jhang T, Singh NK, Koundal KR, Bansal KC, Chandra N *et al*. Genetic differentiation of *Vigna* species by RAPD, URP and SSR markers. Biologia Plantarum. 2007; 51(3):451-457.

29. Dobie P, Haines CP, Hodges RJ, Prevett PF. Insects and arachnids of tropical stored products. Their biology and identification (a training manual). Tropical Development and Research Institute, Berks, UK, 1984, 1-32.

30. Economic survey. Ministry of Finance, Department of Economic affairs. Government of India, 2011-12, 357.

31. Edde PA, Amatobi CI. Seed coat has no value in protecting cowpea seed against attack by *Callosobruchus maculatus* (F.). Journal of Stored Products Research. 2003; 39(1):1-10.

32. Edward YSJT, Gunathilagaraj K. Resistance in Bengal gram (*Cicer orientinum* L.) to the pulse beetle *Callosobruchus maculatus* (F.). (Coleoptera: Bruchidae). Journal of Entomological Research. 1994; 18(3):241-247.

33. Edwards O, Singh KB. Resistance to insect pests: What do legumes have to offer?. Euphytica. 2006; 147:273-285.

34. Eigenbrode SD, Trumble JT. Plant resistance to insects in integrated pest management in vegetables. Journal of Agricultural Entomology. 1994; 11:201-224.

35. Fatokun CA, Menancio-Hautea DI, Dariush D, Young ND. Evidence for orthologous seed weight genes in cowpea and mung bean based on RFLP mapping. Genetics. 1992; 132:841-846.

36. Fordyce JA, Agrawal AA. The role of plant trichomes and caterpillar group size on growth and defense of the pipevine swallowtail *Battus philenor*. Journal of Animal Ecology. 2001; 70:997-1005.

37. Fory LF, Finardi-Filho F, Quintero CM, Osborn TC, Cardona C, Chrispeels MJ *et al*. α-Amylase inhibitors in resistance of common beans to the Mexican bean weevil and the bean weevil (Coleoptera: Bruchidae). Journal of economic entomology. 1996; 89(1):204-210.

38. Fujii K, Miyazaki S. Infestation resistance of wild legumes (*Vigna sublobata*) to azuki bean weevil, *Callosobruchus chinensis* (L.) (Coleoptera: Bruchidae) and its relationship with cytogenetic classification. Applied Entomology and Zoology. 1987; 22:229-230.

39. Gatehouse AMR. Biochemical bases of seed resistance to *Zabrotes subfasciatus* (Bean weevil) in *Phaseolus vulgaris* (common bean): a mechanism for arcelin toxicity. Journal of Insect Physiology. 1990; 36:757-767.

40. Gatehouse JA. Plant resistance towards insect herbivores: a dynamic interaction. New Phytologist. 2002; 156(2):145-169.

41. Giga D. Selection of oviposition sites by cowpea weevils *Callosobruchus rhodesianus* (Pic.) and *Callosobruchus maculatus* (F.). Insect Science Applications. 1995; 16:145-149.

42. Haile A. On-farm storage studies on sorghum and chickpea in Eritrea. African Journal of Biotechnology. 2006; 5:1537-1544.

43. Hartweck LM, Vogelzang RD, Osborn TC. Characterization and comparison of arcelin seed protein variants from common bean. Plant Physiology. 1991; 94:204-211.

44. Horber E. Principles, problems, progress and potential in host resistance to stored grain insects. In: Mills RB, Wright VF, Pedersen JR, McGaughey. Proceedings of the Third International Working Conference on Stored-Product Entomology: Oct, Kansas State University, Manhattan, Kansas/spon. Department of Entomology and Department of Grain Science and Industry, KSU, 1983, 23-28.

45. Horber E. Methods to detect and evaluate resistance in maize to grain insects in the field and in storage. In: Cimmyt, Towards Insect Resistance Maize for the Third World. Proceedings of the International Symposium on Methodologies for Developing Host Plant Resistance to Maize Insects. Cimmyt, 1989.

46. Horber E, Mills RB. Conditions modifying expressing of resistance of maize kernels to maize weevil. Environmental Entomology. 1976; 5:163-168.

47. Hudaib T, Brown S, Goodman AM, Eady PE. Efficacy of artificial seeds in the delivery of bioactive compounds to the seed dwelling larvae of *Callosobruchus maculatus* (Coleoptera: Bruchidae). Arthropod-Plant Interactions. 2013; 7(5):527-533.

48. Humphry ME, Lambrides CJ, Chapman SC, Aitken EAB, Imrie BC. Relationships between hard-seediness and seed weight in mung bean (*Vigna radiata*) assessed by QTL analysis. Plant Breeding. 2005; 124:292-298.

49. Humphry ME, Magner T, McIntyre CL, Aitken EAB, Liu CJ. Identification of a major locus conferring resistance to powdery mildew (*Erysiphe polygoni* DC) in mung bean (*Vigna radiata* L. Wilczek) by QTL analysis. Genome. 2003; 46:738-744.

50. Ishimoto M, Kitamura K. Growth inhibitory effects of an ALPHA-amylase inhibitor from the kidney bean, *Phaseolus vulgaris* (L.) on three species of bruchids (Coleoptera: Bruchidae). Applied Entomology and Zoology. 1989; 24(3):281-286.

51. Jackai LEN, Daoust RA. Insect pests of cowpeas. Annual Review of Entomology. 1986; 31(1):95-119.

52. Kaga A, Ishimoto M. Genetic localization of a bruchid resistance gene and its relationship to insecticidal cyclopeptide alkaloids, the vignatic acids, in mung bean (*Vigna radiata* L. Wilczek). Molecular Genetics and Genomics. 1998; 258:378-384.

53. Kakati P, Deka SC, Kotoki D, Saikia S. Effect of traditional methods of processing on the nutrient contents and some anti-nutritional factors in newly developed cultivars of green gram [*Vigna radiata* (L.) Wilczek] and black gram [*Vigna mungo* (L.) Hopper] of Assam, India. International Food Research Journal, 2010; 17:377-384.

54. Kananji GAD. A study of bruchid resistance and its inheritance in Malawian dry bean germplasm. Ph.D. Thesis. University of Kwazulu Natal, Pietermaritzburg, 2007.

55. Kang MS, Zhang Y, Magari R. Combining ability for maize weevil preference of maize grain. Crop Science. 1995; 35:1556-1559.

56. Kashiwaba K, Tomooka N, Kaga A, Han OK, Vaughan DA. Characterization of resistance to three bruchid species (*Callosobruchus* spp., Coleoptera, Bruchidae) in cultivated rice bean (*Vigna umbellata*). Journal of Economic Entomology. 2003; 96(1):207-213.

57. Keneni G, Bekele E, Getu E, Imtiaz M, Damte T, Mulatu B *et al.* Breeding food legumes for resistance to storage insect pests: potential and limitations. Sustainability. 2011; 3(9):1399-1415.

58. Khajudparn P, Tantasawat P. Relationships and variability of agronomic and physiological characters in mungbean. African Journal of Biotechnology. 2013; 10(49):9992-10000.

59. Kitamura K, Ishimoto M, Sawa M. Inheritance of resistance to infestation with azuki bean weevil in *Vigna sublobata* and successful

incorporation to *V. radiata*. Japanese Journal of Breeding. 1988; 38:459-464

60. Kliebenstein DJ, Lambrix VM, Reichelt M, Gershenzon J, Mitchell-Olds T. Gene duplication in the diversification of secondary metabolism: Tandem 2-oxoglutarate-dependent dioxygenases control glucosinolate biosynthesis in Arabidopsis. The Plant Cell. 2001; 13(3):681-693.

61. Kumar LS. DNA markers in plant improvement: An overview. Biotechnology Advances. 1999; 17(2):143-182.

62. Lakhanpaul S, Chadha S, Bhat KV. Random amplified polymorphic DNA (RAPD) analysis in Indian mung bean (*Vigna radiata* (L.) Wilczek) cultivars. Genetica. 2000; 109(3):227-234.

63. Lale NES, Makoshi MS. Role of chemical characteristics of the seed coat in the resistance of selected cowpea varieties to *Callosobruchus maculatus* (F.) (Coleoptera: Bruchidae) in Nigeria. International Journal of Pest Management. 2000; 46(2):97-102.

64. Lale NES, Kolo AA. Susceptibility of eight genetically improved local cultivars of cowpea to *Callosobruchus maculatus* (F.) (Coleoptera: Bruchidae) in Nigeria, International Journal of Pest Management. 1998; 44:25-27.

65. Lambrides CJ, Godwin ID. Pulses, sugar and tuber crops genome mapping and molecular breeding in plants. Biomedical and Life Sciences. 2007; 3:69-90.

66. Lambrides CJ, Godwin ID, Lawn RJ, Imrie BC. Segregation distortion for seed testa color in mungbean (*Vigna radiata* L. Wilczek). Journal of Heredity. 2004; 95:532-535.

67. Lambrides CJ, Lawn RJ, Godwin ID, Manners J, Imrie BC. Two genetic linkage maps of mung bean using RFLP and RAPD markers. Australian Journal of Agricultural Research. 2000; 51:415-425.

68. Landerito EO, Mendoza EMT, Laurena AC, Garcia RN. Physicochemical and biochemical factors in mungbean (*Vigna radiata*, (L.) Wilczek) and blackgram (*Vigna mungo*) associated with bruchid (*Callosobruchus chinensis* L.) resistance. Philippine Journal of Crop Science. 1993; 18(3):153-163.

69. Laserna-Ruiz I, De-Los-Mozos-Pascual M, Santana-Méridas O, Sanchez-Vioque R, Rodríguez-Conde MF. Screening and selection of

lentil (*Lens miller*) germplasm resistant to seed bruchids (*Bruchus* spp.). Euphytica. 2012; 188(2):153-162.

70. Lattanzio V, Terzano R, Cicco N, Cardinali A, Venere DD, Linsalata V. Seed coat tannins and bruchid resistance in stored cowpea seeds. Journal of the Science of Food and Agriculture. 2005; 85(5):839-846.

71. Li CD, Fatokun CA, Ubi B, Singh BB, Scoles GJ. Determining genetic similarities and relationships among cowpea breeding lines and cultivars by microsatellite markers. Crop Science. 2001; 41(1):189-197.

72. Marconi E, Ruggeri S, Carnovale E. Chemical evaluation of wild under-exploited *Vigna* spp. seeds. Food Chemistry. 1997; 59(2):203-212.

73. McCain FS, Eden WG, Singh DN. A technique for selecting rice weevil resistance in corn in the laboratory. Crop Science. 1964; 4:109-110.

74. Menancio-Hautea D, Fatokun CA, Kumar L, Danesh D, Young ND. Comparative genome analysis of Mung bean (*Vigna radiata* L. Wilczek) and cow pea (*V. unguiculata* L. Walpers) using RFLP mapping data. Theoretical and Applied Genetics. 1993; 86:797-810.

75. Messina FJ, Renwick JAA. Ability of ovipositing seed beetles to discriminate between seeds with differing egg loads. Ecological Entomology. 1985; 10(2):225-230.

76. Narasimhulu R, Naidu NV, Shanthi Priya M, Rajarajeswari V, Reddy KHP. Genetic variability and association studies for yield attributes in mungbean (*Vigna radiata* L. Wilczek). Indian Journal of Plant Sciences. 2013; 2:3.

77. Nwanze KF, Horber E. Laboratory techniques for screening cowpeas for resistance to *Callosobruchus maculatus* F. Environmental Entomology. 1975; 4(3):415-419.

78. Nwanze KF, Horber E. Seed coats of cowpeas affect oviposition and larval development of Callosobruchusmaculatus. Environmental Entomology. 1976; 5(2):213-218.

79. Obiadalla-Ali HA, Salman AMA, El-Hady MAHA. Screening some local and introduced cowpea cultivars for dry-seed yield and resistance to *Callosobruchus maculatus* (F.). Annals of Agricultural Science. 2007; 52:197-212.

80. Osborn TC, Blake T, Gepts P, Bliss FA. Bean arcelin 2. Genetic variation, inheritance and linkage relationships of a novel seed protein of *Phaseolus vulgaris* L. Theoretical and Applied Genetics. 1986;

71:847-855.

81. Painter RH. Insect Resistance in Crop Plants. The Macmillan Company, 1951, 520.

82. Panda N, Khush GS. Host plant resistance to insects. CAB International, Wallingford, 1995.

83. Pedra JHF, Brandt A, Westerman R, Lobo N, Li HM, Romero-Severson J *et al.* Transcriptome analysis ofthe cowpea weevil bruchid: identification of putative protein-ases and alpha-amylases associated with food breakdown. Insect Molecular Biology. 2003; 12:405-412.

84. Rahim MA, Mia AA, Mahmud F, Zeba N, Afrin KS. Genetic Variability, Character Association and Genetic Divergence in Mungbean (*Vigna radiata* L. Wilczek). Plant Omics. 2010; 3(1):1-6.

85. Raina AK. *Callosobruchus* spp. infesting stored pulses (grain legumes) in India and comparative study of their biology. Indian Journal of Entomology. 1970; 32(4):303-310.

86. Raja M, John S. Impact of volatile oils of plants against the cowpea beetle *Callosobruchus maculatus* (FAB.) (Coleoptera: Bruchidae). International Journal of Integrative Biology. 2008; 2(1):62-64.

87. Reddy Amarender A, Reddy GP. Supply Side Constraints in Production of Pulses in India: Case Study of Lentils. Agricultural Economics Research Review. 2010; 23:129-136.

88. Roychowdhury R, Datta S, Gupta P, Jagatpati TAH. Analysis of genetic parameters on mutant populations of mungbean (*Vigna radiata* L.) after ethyl methane sulphonate treatment. Notulae Scientia Biologicae. 2012; 4(1):137-143.

89. Saeed I, Khattak GSS, Zamir R. Assosiation of seed yield and some important morphological traits in mungbean (*Vigna radiata* L.). Pakistan Journal of Botony. 2007; 39(7):2361-2366.

90. Sarwar M. Assessment of resistance to the attack of bean beetle *Callosobruchus maculatus* (Fabricius) in chickpea genotypes on the basis of various parameters during storage. Songklanakarin Journal of Science Technology. 2012; 34(3):287-291.

91. Sarwar M. Studies on resistance discrepancy in seeds of local and exotic lentil (*Lens culinaris* Medikus) genotypes against *Callosobruchus analis* L. (Coleoptera: Bruchidae). African Entomology. 2013; 21(2):273-279.

92. Savelkoul FHMG, Van der Poel AFB, Tamminga S. The presence and inactivation of trypsin inhibitors, tannins, lectins and amylase inhibitors in legume seeds during germination. A review. Plant Foods for Human Nutrition. 1992; 42(1):71-85.

93. Schoonhoven A, Cardona C, Valor J. Resistance to the bean weevil and Mexican bean weevil (Coleoptera: Bruchidae) in non-cultivated common bean accessions. Journal of Economic Entomology. 1983; 76:1255-1259.

94. Semple RL, Hicks PA, Lozare JV, Castermans A. Towards integrated commodity and pest management in grain storage. A training manual for application in humid tropical storage systems. Proceedings and selected papers from the Regional Training Course on Integrated Pest Management Strategies in Grain Storage Systems, National Post Harvest Institute for Research and Extension (NAPHIRE), Department of Agriculture, June, Philippines, 1992, 1988, 6-18.

95. Sen A, Bergvinson D, Miller SS, Atkinson J, Fulcher RG, Arnason JT. Distribution and microchemical detection of phenolic acids, flavonoids, and phenolic acid amides in maize kernels. Journal of Agricultural and Food Chemistry. 1994; 42(9):1879-1883.

96. Shade RE, Kitch LW, Mentzer P, Urdock LL. Selection of a cowpea weevil (Coleptera: Bruchidae) biotype virulent to cowpea weevil resistant landrace TVu 2027. Journal of Economic Entomology. 1996; 89:1325-1331.

97. Shade RE, Murdock LL, Foard DE, Pomeroy MA. Artificial seed system for bioassay of cowpea weevil (Coleoptera: Bruchidae) growth and development. Environmental Entomology. 1986; 15:1286-1291.

98. Shade RE, Schroeder HE, Pueyo JJ, Tabe LL, Murdock TJK, Higgins MJ *et al.* Transgenic pea seeds expression the α-amylases inhibitors of the common bean are resistant to bruchid beetles. Biotechnology. 1994; 12:793-796.

99. Singh S, Sharma G. Screening of chickpea varieties for oviposition preference and larval development of pulse beetle, *Callosobruchus chinensis* (Linn.). Pest Management and Economic Zoology. 2001; 9(1):19-22.

100. Singh S, Sharma G. Pulse beetle management through solar energy in stored green gram seeds. Seed Research, New Delhi. 2003; 31(1):84-89.

101. Singh SR, Jackai LEN. Insect pests of cowpea in Africa: their life cycle, economic importance and potential for control. In: Singh SR, Rachiek, O. (Eds.), Cowpea Research, Production and Utilisation. Wiley Interscience Publications/John Wiley and Sons Ltd, Chichester, 1985, 217-231.

102. Somta C, Somta P, Tomooka N, Ooi PAC, Vaughan DA, Srinives P. Characterization of new sources of mung bean (*Vigna radiata* (L.) Wilczek) resistance to bruchids, *Callosobruchus* spp. (Coleoptera: Bruchidae). Journal of Stored Products Research. 2008; 44:316-321.

103. Southgate BJ. Systematic Notes on Species of *Callosobruchus* of economic importance. Bulletin of Entomological Research. 1958; 49:591-599.

104. Srivastava AC, Tiwari LD, Madan Paland UK, Sengupta. CO2-mediated changes in mung bean chemistry: Impact on plant-herbivore interactions. Current Science. 2002; 82:9.

105. Suzuki K, Ishimoto M, Iwanaga M, Kikuchi F, Kitamura K. Inheritance of seed α- amylase inhibitor in the common bean and genetic relationship to arcelin. Theoretical and Applied Genetics. 1995; 90:762-766.

106. Talekar NS. Biology, damage and control of bruchid pests of mung bean, in: Mung bean: *Proceedings of the 2nd International Symposium*. Edited by S. Shanmugasundaram and BT McLean. AVRDC, Tainan, Taiwan, 1988, 329-342.

107. Talekar NS, Lin CP. Characterization of *Callosobruchus chinensis* (Coleoptera: Bruchidae) resistance in mungbean. Journal of Economic Entomology. 1992; 85(4):1150-1153.

108. Tan H, Tie M, Luo Q, Zhu Y, Lai J, Li H. A review of molecular markers applied in cowpea (*Vigna unguiculata* L. Walp.) breeding. Journal of Life Science. 2012; 6:1190-1199.

109. Tantasawat P, Trongchuen J, Prajongjai T, Thongpae T, Petkhum C, Seehalak W *et al.* Variety identification and genetic relationships of mung bean and black gram in Thailand based on morphological characters and ISSR analysis. African Journal of Biotechnology. 2010; 9(27):4452-4464.

110. Taylor TA. Distribution, ecology and importance of bruchids attacking grain legumes and pulses in Africa. In: Labeyrie, V. (Ed.). The Ecology

of Bruchids Attacking Legumes (pulses). Proceedings of the International Symposium, Tours, France, Junk Publishers, The Hague, Netherlands, 1981.

111.Tipping PW, Cornelius PL, Legg DE. Inheritance of resistance in whole kernel maize to oviposition by the maize weevil (Coleoptera: Curculionidae). Journal of Economic Entomology. 1989; 82:1466-1469.

112.Udayagiri S, Wadhi SR. Catalog of Bruchidae. Memoirs of the American Entomological Institute. 1989; 45:301.

113.Van Huis A, De Rooy M. The effect of leguminous plant species on *Callosobruchus maculatus* (Coleoptera: Bruchidae) and its egg parasitoid *Uscana lariophaga* (Hymenoptera: Trichogrammatidae). Bulletin of Entomological Research. 1998; 88(1):93-99.

114.Wang XW, Kaga A, Tomooka N, Vaughan DA. The development of SSR markers by a new method in plants and their application to gene flow studies in azuki bean [*Vigna angularis* (Willd.) Ohwi & Ohashi]. Theoretical and Applied Genetics, 2004; 109(2):352-360.

115.Wolie A, Dessalegn T, Belete K. Heritability, variance components and genetic advance of some yield and yield related traits in Ethiopian collections of finger millet (*Eleusine coracana* (L.) Gaertn.) Genotypes. African Journal of Biotechnology. 2013; 12(36):5529-5534.

116.Young ND, Danesh D, Menancio-Hautea D, Kumar L. Mapping Oligogenic resistance to powdery mildew in mung bean with RFLPs. Theoretical and Applied Genetics. 1993; 87:243-249.

117.Zaid IU, Khalil IH, Khan S. Genetic varibaility and correlation analysis for yield components in mung bean (*Vigna radiata* L. Wilczek). ARPN Journal of Agricultural and Biological Science. 2012; 7(11):885-991.

Chapter - 8
Biological Control of Brinjal Shoot and Fruit Borer, *Leucinodes orbonalis* Guenee

Authors

Selvi C
ICAR-Central Institute for Cotton Research, Regional Station, Coimbatore, Tamil Nadu, India

Arunkumar N
Department of Agricultural Microbiology, TNAU, Coimbatore, Tamil Nadu, India

Guru Prasanna Pandi G
Division of Crop Protection, ICAR-National Rice Research Institute, Cuttack, Odisha, India

Soumia P.S
Division of Crop Protection, ICAR-Directorate of Onion and Garlic Research, Pune, Maharashtra, India

Sivakumar G
Division of Insect Ecology, ICAR-NBAIR, Bangalore, Karnataka, India

Chapter - 8

Biological Control of Brinjal Shoot and Fruit Borer, *Leucinodes orbonalis* Guenee

Selvi C, Arunkumar N, Guru Prasanna Pandi G, Soumia P.S and Sivakumar G

1. Introduction

Brinjal, *Solanum melongena* L., is an important solanaceous vegetable crop and is extensively cultivated in South Asia. Among biotic factors, Brinjal shoot and fruit borer (BSFB) (*Leucinodes orbonalis* Guenee) is considered as a major threat limiting the production of brinjal. For the management of BSFB several management tactics have been formulated and recommended .Farmers mainly rely on insecticides to manage this devastating pest. The indiscriminate and injudicious application of synthetic insecticides has given rise to the problems such as residual toxicity, pesticide resistance, resurgence, secondary pest outbreak, potential health and environmental threats and destruction of natural enemies of pest. To overcome the problems associated with chemical pesticides biological control would be a better alternative. For the management of BSFB several biocontrol approaches such as use of parasitoids, and predators, entomopathogens such as bacteria, virus, fungi and nematodes have been documented both under laboratory and field conditions by several workers. Among the biocontrol agents the egg parasitoid, *Trichogramma chilonis* Ishii have been demonstrated to be most effective under field condition. Apart from parasitoids, entomopathogenic nematode, *Steinernema carpocapsae* that are good candidates for the biological control of BSFB, either alone or in conjunction with other pest management systems.

Brinjal is an important cash crop for more than 1.4 million small, marginal and resource-poor farmers and grown in almost all states of India. According to the *National Horticulture Board, brinjal* was grown over six lakh hectares in 2009, with an annual production of 10. 378 m tonnes in India. In Tamil Nadu, brinjal was cultivated in 7,200 hectares with an annual production of 80,000 metric tonnes. It is a good source of vitamin A, B and C and minerals especially iron, phosphorus and calcium (Patra *et al.*, 2009). Due to its low calorific value (24 kcal 100 g^{-1}) and high potassium content

(200 mg 100 g^{-1}), it is suitable for diabetes, hypertensive and obese patients (Prabhu *et al.*, 2009). Besides being used as an important vegetable, brinjal has been exploited extensively in traditional medicines (Khan, 1979).

Brinjal is attacked by a number of insect pests, diseases and nematodes during various stages of crop growth in most of the tropical countries including India. This cash earning crop is damaged by more than 36 insect pests, from the nursery stage to harvest (Reghupathy *et al.*, 1997). Brinjal crop during its different growth stages is attacked by a number of insect pests *viz.*, aphids, *Aphis gossypii* Glover; leaf hopper, *Amrasca bigutulla bigutulla* (Ishida); Whitefly, *Bemisia tabaci* Genn., mealy bug *Coccidohystrix insolitus* Green., two-spotted spider mite, *Tetranychus urticae* Koch., thrips *Scirtothrips dorsalis* (Hood) brinjal stem borer, *Euzophera perticella* Rag., hadda beetles, *Epilachna dodecastigma* Mulsant and *Epilachna vigintioctopunctata* Fabricius, ash weevil *Myllocerus* spp. leaf folder, *Eublemma olivacea* (Walker) and shoot and fruit borer (BSFB), *Leucinodes orbonalis* Guenee (Pyralidae: Lepidoptera). Among various pests attacking brinjal shoot and fruit borer (BSFB) *L. orbonalis*, has been found to be the most serious and destructive one. It inflicts damage to both shoots and fruits (Srinivasan, 2008; Arora, 2009). It has been reported that the insect caused a yield loss up to 70 per cent (Nair, 1986; Ghosh *et al.*, 2003; Dhandapani *et al.*, 2003).

2. Biology and Symptoms of Damage of BSFB

The adult is a grey-brown moth with white wings marked brown. Eggs are laid singly on leaves, shoots and fruits. They hatch in three to five days. On hatching, the larvae crawl for about 30-60 minutes looking for a suitable site for penetration and bore inside the top shoot and tender fruits. Larvae undergoes five instars and becomes full fed in 10-15 days when it is stout and pink coloured measuring about 1.6 cm in length. Pupation of the larva takes place in a cocoon on the plant debris for two to five days and the female lays up to 250 eggs (Fig.1.). The BSFB starts damaging brinjal plant a few weeks after the transplanting. When the shoot is attacked by the caterpillar it droops and withers and finally dries up. When the petioles of the leaves are bored by the larvae the leaves wither and droop. The attacked fruits show holes on them plugged with excreta which make the fruit unmarketable (Nair, 1986).

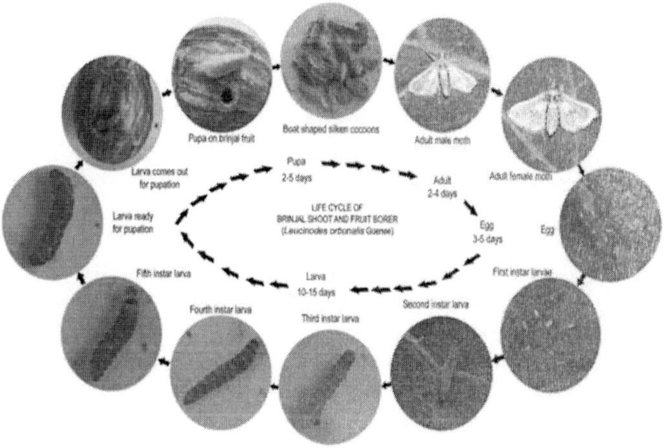

Fig 1: Life cycle of brinjal shoot and fruit borer (BSFB), *Leucinodes orbonalis*

Despite serious nature of the pest, its management by and large is limited to frequent sprays of chemical insecticides to counter the threat of BSFB. Since the larvae are concealed within shoots and fruits, the pest can easily escapes from insecticide sprays. Therefore farmers tend to over-spray insecticides, as they rely mainly on the subjective assessments of the visual presence of the pest. In India, brinjal growers spray pesticides on an average of 27 times over a period of eight months and most of these sprays were targeted against BSFB (Kolady and Lesser, 2006). In addition to the financial cost associated with indiscriminate insecticide applications and its negative effects on the environment, high pesticide residues in brinjal fruits pose serious risk to consumer's health and safety. Management of BSFB by biological control is an alternative strategy that results in pesticide-free produce with no hazard to the human health and environment. Under natural conditions, natural enemies keep the pest populations under reasonable control.

3. Biological Control

Biological control involves the use of natural enemies (parasitoids, predators and pathogens) that are antagonists of insects. As many as sixteen parasitoids, three predators and three species of entomopathogens have been reported as natural enemies of *L. orbonalis* from all over the world (Khorsheduzzaman *et al.*, 1998).

3.1 Parasitoids

More than a dozen parasitoids have been recorded as natural enemies of

L. orbonalis (Khorsheduzzaman *et al*., 1998). Of particular interest, parasitoids of the genus *Trichogramma*, which are egg-parasitoids in the family Trichogrammatidae, have been reported to be efficient in regulating populations of BSFB in brinjal Ecosystems. But the extent of parasitism under field condition is very low. In this circumstance, inundative release of biocontrol agents, mainly egg parasitoids in pest management may be more beneficial. Among them, the egg parasitoid *Trichogramma chilonis* Ishii have been demonstrated to be most effective (Krishnamoorthy, 2012) and it was slightly more efficient than *Trichogramma evanescens* Westwood in parasitizing the egg in test tube and in caged conditions. *T. chilonis* 1 to 1.5 lakh/ha was found effective for the management of BSFB as ecofriendly control (Sathe *et al.,* 2016). Hence, it can be further integrated with other pest management tactics to increase the yield (Gonzales, 1999). The efficacy of *T. chilonis* may further be enhanced by improving the release technique and better integration with other tactics (Satpathy *et al*., 2005). Recently, Niranjana *et al*. (2015) studied the parasitizing efficiency and emergence rate of the egg parasitoids *T.chilonis* and *Trichogramma pretiosum* and results demonstrated that *T. pretiosum* was also an efficient egg parasitoid against the eggs of *L. orbonalis* was studied under laboratory condition. However, this parasitoid was highly vulnerable to the temperature existed in the summer season. *T. pretiosum* has the potential to parasitize the eggs of *L. orbonalis* even in the absence of male insects by parthenogenesis.

Bustamante *et al.* (1994) recorded the egg parasitoid *T. chilonis* Ishii is most effective against *L. orbonalis* that contributed highest yield followed by insecticides such as Endosulfan and Endosulfan + Deltamethrin, botanical and fungus *Metarhizium anisopliae* (Metchnikoff), respectively. Sandanayake and Edirisinghe (1992) reported a high level of parasitism of *L. orbonalis* larvae by a parasitoid, *Trathala flavo-orbitalis* (Cameron) (Hymenoptera: Ichneumonidae). This parasitoid has been reported to be present in India (Naresh *et al.,* 1986) and Bangladesh (Alam and Sana, 1964), Nepal (Kafle 1970) however; its contribution to pest control has hardly been documented and does not appear to be significant. However, Srinivasan (2008) reported that practice of withholding of pesticide use to allow proliferation of local natural enemies especially the parasitoid, *T. flavoorbitalis* reduced the *L. orbonalis* population.

Endo-parasitoids, such as *Camptothlipsis* sp., *Campyloneurus mutator* Fabricius, *Chelonus* sp. and *Cremastus* (*Trathala*) *flavoorbitalis* (Cameron) (Hymenoptera) were documented as natural enemies of the *L. orbonalisi* (Kafle, 1970). Under natural condition, the parasitism of *L. orbonalis* by hymenopterous parasitods was recorded as an average of 23.33 percent during May-June (Ghimire *et al.,* 2001).

In Tamil Nadu, Yasodha and Natarajan (2009) documented the parasitoids such as *Trathala flavoorbitalis* Cameroon, *Phaneratoma* sp. *Chelonus* sp., *Vaepellinae* sp. *Bracon hebetor* Say, *Antrocephalus mitys* Walker, *Brachymeria lasus* Walker, *Spalangia irregularis* Walker, *Spalangia endius* Walker, *Endius* sp. *Spalangia* sp. and *Trichogramma* sp. *Trichogramma* sp. was found to parasitize the field exposed eggs of *L. orbonalis* for six days.

Several workers recorded various groups of natural enemies on *L. orbonalis* viz., The ichneumonids *Peristomerus testaceus*, *Eriborus argentiopilosus (*Tewari, and Singh 1987), *Bracon chinensis* Nair, R.K., 1967, *Apanteles* spp., *Chelonus* spp, (Gokulpur, R.S., 1970; Navasero, and Calilung,1990 and the chalcids *Brachymeria obscurata* (Walker). Sathe *et al.* (2016) parasitoids of BSFB such as *Trathala flavoorbitalis, Prestomerus testaceus, Bracon sp.Shirakia shoenobi, Diadegma apastata, Eriborus argenteopilosus, Trichogramma chilonis* have been recorded from Kolhapur region of India.

3.2 Predators

Coccinellid beetles such as *Coccinella septumpuctata, Cheilomenes sexmaculatus Brumoides suturalis* identified as predators of BSFB (Sathe *et al.,* 2016). There is also significant relationship between incidence of *L. orbonalis* in terms of shoot infestation and with coccinellids and spiders (Singh *et al.,* 2009).

4. Entomopathogens as Pest Control Agents

Bio-pesticides have high compatibility with other pest management techniques, such as natural enemies, resistant varieties, etc. Therefore, it could play a vital role in integrated pest management (IPM) strategies although they cover only about 4 percent of the global pesticide market (Srinivasan, 2012). They have minimal effect on non- target organisms. Similarly, the potential human health effects are not known so far. The combination of plant products and synthetic microbial pesticides are reported to possess insect controlling properties. These biopesticides can provide environmentally safe product for bio-rational management of brinjal pest.

4.1 Bacteria

Among the bacteria, the spore forming species, *Bacillus* spp. have been most important in pest management. Formulations based on the insect pathogenic bacterium *B.thuringiensis* (Berliner) occupy the key position of the worlds leading bio pesticide accounting for an average of 90 per cent of bio pesticide sales (Neale, 1997). It has been in limited field use over a period of 40 years for Lepidopteran pests control. Different commercial *B. thuringiensis* formulations (Delfin 25WG, Dipel 8L, Halt WP and Biolep WP, Biobit WP, Spicturin) were evaluated either alone or in comparison with neem and chemical insecticides against BSFB. These sprays were reported to be effective against *L. orbonalis* under field conditions and resulted in minimum shoot (9.56%) as well as fruit (11.78%) infestation (Abilash, 2000; Puranik *et al.*, 2002; Yashodha and Kuppusamy, 2008).

Saimandir and Gopal (2012) showed that Biolep was not effective; however, PUSA Bt gave better control against *L. orbonalis* at similar dose. Chatterjee and Mondal (2012) reported that *B.thuringiensis* performed well in reducing damage and increasing yield. BT formulation Biolep was found to be effective than synthetic insecticide (Sharma, 2002) and efficacy increased in successive days of spray (Mainali *et al.*, 2012). However, the efficacy of BT formulation was lower or comparative to pesticides as reported by Singh (2010) that the marketable yield from BT formulation halt was comparatively lower but as efficient as Spinosad, Emamectin Benzoate, Diafenthiuron, Endosulfan and Indoxacarb. Different commercial *B. thuringiensis* formulations (Delfin 25WG, Dipel 8L, Halt WP and Biolep WP, Biobit WP, Spicturin) were evaluated either alone or in comparison with neem and chemical insecticides against BSFB. Results of pathogenicity test of three entomopathogenic fungi@ 2 x 10^9 conidia per ml and *B.t. kurstaki* showed maximum mortality by *Verticillium lecanii* against all larval instars of *L. orbonalis* (Niranjana, 2015)

4.2 Virus

Similarly, to date more than 450 viruses have been described. The majority of these cause diseases in Lepidoptera, Hymenoptera and Diptera. The best-known virus of insect is the Nuclear Polyhydrosis Viruses (NPV). Under field condition, even lepidopteran larvae are very sensitive to these biocides (Gupta and Rosan, 1995). However, the efficacy NPV against *L. orbonalis* is lower but it can be used as one of the important options of management (Ghimire, 2001).

4.3 Entomopathogenic Fungi

Literatures regarding efficacy of fungal pesticides against brinjal shoot and fruit borer of brinjal are very limited, Efficacy of the fungal pathogens, *Nomuraea rileyi, Metarhizium anisopliae and Beauveria bassiana* were studied in the management of BSFB (Koushik *et al.*, 2012; Tripura *et al.*, 2017). The influence of different concentrations of *N. rileyi* on the mortality of larvae of *L. orbonalis* was studied in laboratory condition. Combination *Fusarium semitectum* $3.6x10^{15}$+ *N. rileyi* $2x10^{8}$+ spinosad (15 g a.i/ha) with insecticides and other biocontrol agents *N. rileyi was* effective in suppression of fruit borer population. *M. anisoplae, B. thuringiensis* and *B. bassiana* were not so effective treatments (Tripura *et al.,* 2017). Studies were conducted to evaluate the efficacy of plant products *viz.*, neem oil (2%), iluppai oil (2%), pungam oil (2%), combination of iluppai and pungam (1: 1) and microbial formulations *viz.*, entomopathogenic fungi, *B. bassiana* and *Verticillium lecanii* against *L. orbonalis.* Thus fungal based biopesticides alone was not much effective against BSFB. However, in cabining with other pest management efficacy was recorded by several earlier workers.

4.4 Entomopathogenic nematodes (EPN)

Among the different agents for biological control, entomopathogenic nematodes are gaining importance, because they possess many positive attributes of an effective biological control agent. EPN often have broad-spectrum effectiveness, short life cycles, amenability to mass production, recycling ability, persistence etc. (Kaya and Gaugler, 1993). Field efficacy of two common isolates, viz., POBC EN 6.11 of *Steinernema carpocapsae* and POBC EN 13.3 of *Heterorhahditis indica* was evaluated against *L. orbonalis* in brinjal at experimental fields of Indian Institute of Horticultural Research, Bangalore in 2002. Results indicated that among the two strains of EPN S. *carpocapsae* was more effective in reducing the fruit damage by BSFB. In vitro studies carried out by Subramanian (2000) and Hussaini *et al.* (2002) showed that *Steinenerma* spp. are potential biological control agents of brinjal shoot and fruit borer. Field trials were also conducted for three consecutive years (2005-07) to assess the efficacy of an isolate of the *S.carpocapsae* against this pest under field conditions caused significant reduction in fruit borer damage and increased yield (Ganga Visalakshy *et al., 2009),*

5. Conclusion

Application of insecticides to combat *L. orbonalis* often results in unsatisfactory control. Hence, farmers are willing to adopt alternative control measures such as biological control. This measure is safe and effective, although the cost of the treatments may be greater than that of chemical pesticides. Scores of parasitoids and entomopathogenic nematodes (Rahoo *et al.*, 2017) have been reported as natural enemies of ESFB throughout the world and are responsible for keeping the pest populations under reasonable control. Among these, the egg parasitoid *T. chilonis* and entomopathogenic nematode, *S. carpocapsae have* been proved to be the most effective and caused increase in yield. However, timely availability of good quality bio-control agents is one of the factors that are presently hindering biological control in India. The biocontrol agent can also be integrated with other pest management tactics to increase crop yield.

References

1. Abilash RBD. Studies on development of IPM techniques for brinjal shoot and fruit borer, *Leucinodes orbonalis* Guenee. M.Sc. (Ag.) Thesis, Tamil Nadu Agricultural University, Coimbatore, India, 2000, 93.
2. Alam AZ, Sana DL. Biology of *Leucinodes orbonalis* Guen. In: Alam MZ and Sattar A (Eds.). Review of research in East Pakistan, Division of Entomology (1947-1964): Agricultural Information Service and East Pakistan. Agriculture Research Institute Dhaka, 1964, 192-200.
3. Arora S. Analysis of insecticides in okra and brinjal from IPM and non-IPM fields. Environ. Monit. Assess. 2009; 151:311-315.
4. Bustamante RC, Luzaran PB, Gruber LT, Villaneva E. Field evaluation on different control methods against brinjal fruit and shoot borer. The Philippine Journal of Plant Industry. 1994; 59(1-2):119-125.
5. Chatterjee ML, Mondal S. Sustainable management of key lepidopteran insect pests of vegetables. ISHS Acta Horticulturae. (Available at: http://www.actahort.or g/books/958/ 958_17.htm Retrieved on April 4, 2013), 2012.
6. Dhandapani N, Shelkar UR, Murugan M. Bio-intensive pest management in major vegetable crops: An Indian perspective. J Food Agr. Environ. 2003; 1(2):330-339.
7. Ganga Visalakshy PN, Krishnamoorthy A, Hussaini SS. Field efficacy of the entomopathogenic nematode *Steinernema carpocapsae* (weiser, 1955) against brinjal shoot and fruit borer, *Leucinodes orbonalis* guenee. Nematol. medit. 2009; 37:133-137.
8. Ghimire A, Khatiwada BP. Use of pesticides in commercial vegetable cultivation in Tandi, Eastern Chitwan, Nepal. Survey report submitted to Department of Entomology Institute of Agricultural and Animal Science (IAAS), Rampur, Chitwan, Nepal, 2001, 10.
9. Ghimire SN. Eco-frendly management of brinjal fruit and shoot borer,

Leucinodes orbonalis Guenee (Lepidoptera: Pyralidae). M.Sc. Ag. Thesis, Department of Horticulture, IAAS, Rampur, Chitwan, Nepal, 2001, 80.

10. Ghosh SK, Laskar N, Senapati SK. Estimation of loss in yield of brinjal due to pest complex under Terai region of West Bengal. Environ. Ecol. 2003; 21(4):764-769.

11. Gokulpur RS. Records of braconids and ichneumonids from Cen- tral India. Indian J ENT. 1970; 32(4):383-385.

12. Gonzales PG. Management of eggplant shoot and fruit borer, *Leucinodes orbonalis* Guenee. M. S. Thesis, University of Los Banos, Laguna, Philippines, 1999, 87. (Abs.).

13. Gupta GR, Rosan L. Biopesticide as alternative to synthetic insecticide for pest management in cotton. In: Walia S and parmer BS (Eds). Pesticide Crop Protection and Environment. Oxford and IBH Publishing co. Pvt. Ld. New Dehli, India, 1995, 454-465.

14. Hussaini SS, Singh SP, Nagesh M. *In vitro* and field evaluation of some indigenous isolates of *Steinernema* and *Heterorhabditis indica* against shoot and fruit borer, *Leucinodes orbonalis*. Indian Journal of Nematology. 2002; 32:63-65.

15. IIHR. *Trichogramma chilonis* effciency against *Leucinodes orbonalis* (Guinee). ICAR, New Delhi, India, 2004, 128. (Annual report).

16. Kafle GP. New records of some insects and non-insects parasites of Leucinodes orbonalis Guenee in Nepal. Nepalese. Journal of Agriculture. 1970; 5:99-103.

17. Kaya H, Gaugler R. Entomopathogenic nematodes. Annual Review of Entomology. 1993; 38:181-206.

18. Khan R. *Solanum melongena* and its ancestral forms. In: The biology and taxonomy of Solanaceae. J Hawkes RN, Lester and AD Skelding. (eds) Academic, London, 1979, 629-636.

19. Khorsheduzzaman AKM, MI Ali, MA Mannanm, Ahmad A, Karim MA. Component technologies of IPM package for brinjal shoot and fruit borer for Bangladesh: A review. Bangladesh Journal of Agricultural Research. 1998; 23:593-606.

20. Kolady DE, Lesser W. Who adopts what kind of technologies? The case of *Bt* Eggplant in India. Ag Bio Forum. 2006; 9:94-103.

21. Koushik NR, Manjunatha M, Naik MI, Shivanna BK, Narayanaswamy H, Kalleswara Swamy CM *et al*. Evaluation of eco-friendly approaches against brinjal shoot and fruit borer *Leucinodes orbonalis* (Guenee) (Lepidoptera: Pyralidae). Karnataka J Agric. Sci. 2013; 26(2):209-213.

22. Krishnamoorthy A. Exploitation of egg parasitoids for control of potential pests in vegetable ecosystems in India. Comunicata Scientiae. 2012; 3(1):1-15.

23. Mainali RP. Bio-rational management of eggplant fruit and shoot borer, *Leucinodes orbonalis* Guenee, (Lepidoptera: Pyralidae) in Nepal. M. Sc. Ag. Thesis. Institute of Agriculture and Animal Sciences/TU, Chhimireitwan, Nepal, 2012, 109.

24. Mathur AS, Singh N, Singh P, Meena M. Field evaluation of plant products and microbial formulations against brinjal shoot and fruit borer, *Leucinodes orbonalis* Guenee under semi-arid conditions of Rajasthan. J Biopest. 2012; 5(1):71-74.

25. Nair MRGK. Insects and mites of crops of India. Indian Council of Agricultural Research, New Delhi, India, 1986, 151-152.

26. Nair RK. *Leucinodes orbonalis* as a serious pest of potato at two places in Mysore state. Indian J ENT. 1967; 29(1):96-97.

27. Naresh JS, Malik VS, Balan JS. Estimation of fruit damage and larval population of brinjal fruit borer, *Leucinodes orbonalis* Guen. And its parasitization by *Trathala* sp. on brinjal. Bulletin of Entomology (India). 1986; 27:44-47.

28. Navasero MV, Calilung VJ. Biology of the egg plant fruit and shoot borer, *L. orbonalis* in Philippines. Philippine Entomologist. 1990; 8(2):749-760.

29. Neale MC. Bio-pesticides-harmonisation of registration requirements within EU directive. An industry view. Bulletin of European and Mediterranean Plant Protection Organization. 1997; 27:89-93, 91-414.

30. Niranjana RF, Devi1 M, Shanika W, Philip Sridhar R. Potential Use of Egg Parasitoids, *Trichogramma pretiosum* Riley and *Trichogramma chilonis* Ishii Against Brinjal Shoot and Fruit Borer, *Leucinodes orbonalis* Guenee. Tropical Agricultural Research. 2015; 27(1):88-94.

31. Patra S, Mondal MLCS, Samanta A. Field evaluation of some new insecticides against brinjal shoot and fruit borer *Leucinodes orbonalis* Guenee. Pestic. Res. J, 2009; 21(1):58-60.

32. Prabhu M, Natarajan S, Veeraragavathatham D, Pugalendhi L. The biochemical basis of shoot and fruit borer resistance in interspecific progenies of brinjal (*Solanum melongena*). Eur. Asia J Bio. Sci. 2009; 3:50-57.

33. Reghupathy A, Palanisamy S, Chandramohan N, Gunathilagaraj K. A guide on crop pests. Sooriya Desktop Publishers, Coimbatore, 1997, 264.

34. Saimandir J, Gopal M. Evaluation of synthetic and natural insecticides for the management of insect pest control of e Saimandir ggplant (*Solanum melongena* L.) and Pesticide Residue Dissipation Pattern. American Journal of Plant Sciences. 2012; 3(2):214-227.

35. Sandanayake WRM, Edirisinghe JP. *Trathala flavoorbitalis*: parasitization and development in relation to host-stage attacked. Insect Science and Its Application. 1992; 13(3):287-292.

36. Sathe TV, Patil SS, Bhosale AM, Devkar SS, Govali CS, Hankare SS. Ecology and Control of Brinjal insect pests from Kolhapur region, India. Biolife. 2016; 4(1):147-154.

37. Satpathy S, Shivalingaswamy TM, Kumar A, Rai AB, Rai M. Bio intensive management of eggplant shoot and fruit borer (*Leucinodes orbonalis* guen.) Veg. Sci. 2005; 32(1):103-104.

38. Sharma DR. Bioefficacy of certain insecticide and biopesticides against major pest of brinjal under field condition. M. Sc. Thesis, Indian Agriculture Research Institute, New Delhi, India, 2002, 160.

39. Singh DK, Singh R, Datta SD, Singh SK. Seasonal incidence and insecticidal management of shoot and fruit borer (*Leucinodes orbonalis* Guenee) in brinjal. Annals of Horticulture. 2009; 2(2):187-190.

40. Singh SS. Comparative efficacy of certain bio-rational insecticides and *Bacillus thuringiensis* (Berliner) basedbio-insecticides against *Leucinodes orbonalis* Guen. In brinjal. Indian Journal of Horticulture. 2010; 67(3):353-356.

41. Srinivasan R. Integrated Pest Management for eggplant fruitand shoot borer (*Leucinodes orbonalis* Guenee) in south and Southeast Asia: past, present and future. Journal of Biopesticides. 2008; 1(2):105-112.

42. Srinivasan R. Integrating bio-pesticides in pest management strategies for tropical vegetable production. Journal of Biopesticides. 2012; 5:36-45.

43. Subramanian S. Studies on the entomopathogenic nematodes *Heterorhabditis indica* Poinar, Karunakar and David and *Steinernema glaseri* (Steiner). Ph.D thesis in plant Nematology, TNAU, Coimbatore, India, 2000, 113.

44. Tewari GC, Singh SJ. New record of baculovirus disease in *L. orbonalis* Guen. Curr. Sci. 1987; 56(13):671-672.

45. Tripura A, Chatterjee ML, Pande R, Patra S. Biorational management of brinjal shoot and fruit borer (*Leucinodes orbonalis* guenee) in mid hills of Meghalaya. Journal of Entomology and Zoology Studies. 2017; 5(4):41-45.

46. Yashodha P, Kuppusamy C. A bioefficacy of *Bacillus thuringiensi* var.*kurstaki* against eggplant borer *Leucinodes orbonalis* Guenee (Pyraustidae: Lepidoptera), Int. J Integ. Biol. 2008; 3(3):209-213.

47. Yasodha PN. Natarajan. Record of parasitoids and pathogens on *Leucinodes orbonalis*. Madras Agric. J, 2009; 96(1-6):194-199.

Chapter - 9

Plant Disease Detection-Traditional Methods vs Modern Methods

Authors

Thava Prakasa Pandian R

ICAR-Central Plantation Crops Research Institute Regional Station, Vittal, Karnataka, India

Priyanka S

Department of Plant Breeding and Genetics, Tamil Nadu Agricultural University, Coimbatore, Tamil Nadu, India

Arunkumar N

Department of Agricultural Microbiology, Tamil Nadu Agricultural University, Coimbatore, Tamil Nadu, India

C. Selvi

Tamil Nadu Agricultural University, Coimbatore, Tamil Nadu, India

Chapter - 9

Plant Disease Detection-Traditional Methods vs Modern Methods

Thava Prakasa Pandian R, Priyanka S, Arunkumar N and C. Selvi

Introduction

Plant diseases are responsible for major economic losses in the agricultural industry worldwide. Monitoring plant health and detecting pathogen early are essential to reduce disease spread and facilitate effective management practices. Correct identification of the pathogen may results in proper management of the plant disease. In addition to the visual observation for symptoms, serology and nucleic acid based methods made a great contribution in disease diagnosis. Every methodology has its own pros and cons which will be briefed in this summary. Also novel approaches developed and/or under development for the management of plant diseases are discussed here in brief. One of the main lacunas in traditional methods of disease diagnosis is less reliable during hidden or asymptomatic stage. In the previous chapter 2 we have discussed in detail about the recent trends in detection and diagnosis of plant diseases. Here we have discussed the advantages of the modern methods of plant disease detection with those of traditional counterpart.

Traditional methods of Plant disease detection

Traditionally, plant diseases are detected based on symptomatology (visual inspection), serology (Antigen-Antibody based interaction E.g. ELISA) and nucleic acid (E.g. PCR). But these methods can detect the pathogen only after the appearance of symptoms.

Visual Inspection

Visual symptoms are important for diagnosis of plant diseases. A symptom of plant disease is a visible effect of disease on the plant (Figure 1). Symptoms may include a detectable change in color, shape or function of the plant as it responds to the pathogen. Leaf wilting is a typical symptom of verticilium wilt, caused by the fungal plant pathogens *Verticillium alboatrum* and *V. dahliae*.

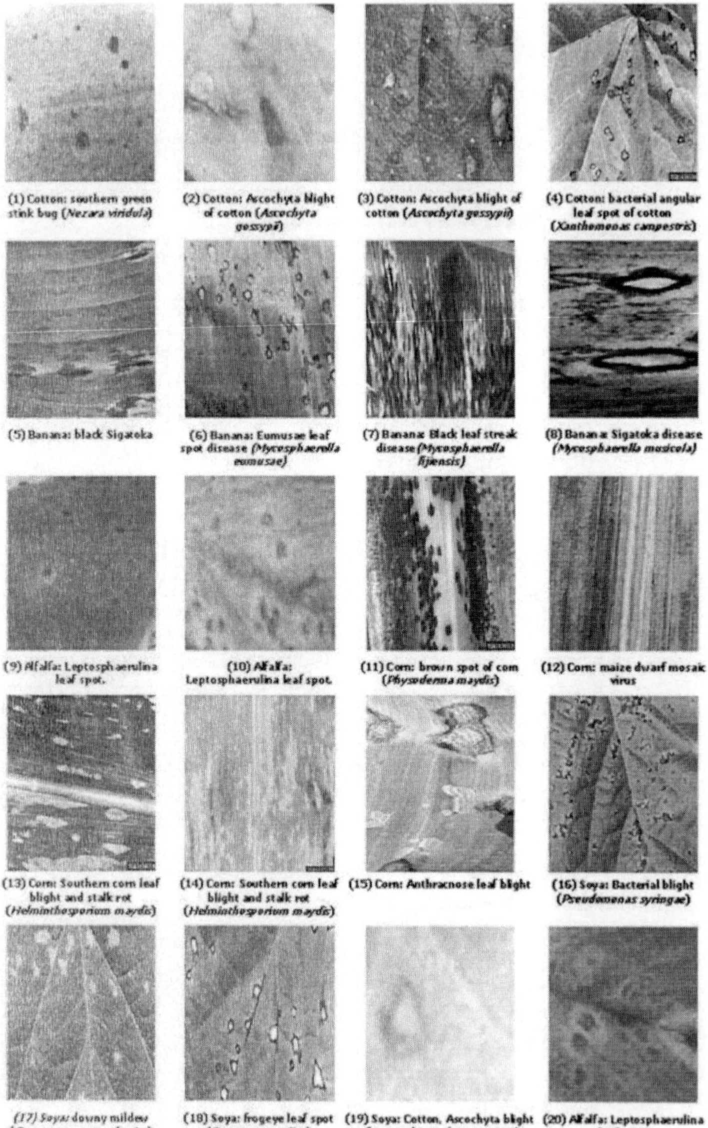

Fig 1: Important plant diseases and their visual symptoms (Camargoa and Smith, 2009)

Common bacterial blight symptoms include brown, necrotic lesions surrounded by a bright yellow halo at the leaf margin or interior of the leaf on bean plants. During the past years, plant diseases were diagnosed with accuracy and precision using traditional disease scales based on the visual symptoms. Nevertheless, these methods are too biased. Sometimes it may lead to wrong identification of diseases which led us to take up wrong management practices.

Serology Based Detection Methods

Serology based techniques such as enzyme-linked immuno-sorbent assay (ELISA) (Figure 2), western blots, immune strip assays, dot-blot immune-binding assays, and serologically specific electron microscopy (SSEM) which used polyclonal and monoclonal antisera for the detection of many plant pathogenic viruses, bacteria and fungi. Though ELISA is by far the most widely used and potential serological assay, polyclonal antisera (can detect multiple epitopes on an individual antigen) sometimes resulted in cross-reactivity which may further lead to wrong identification. Moreover their shelf life is also short. To address these monoclonal antibodies (MCAbs) were developed using 'hybridoma technology' (Kohler and Milstein, 1975) which detect one epitome only; which are more expensive. Irrespective of these cons, ELISA procedures using both poly- and monoclonal antibodies are available for numerous taxa of plant pathogenic bacteria; similarly, rapid detection kits are also commercially available.

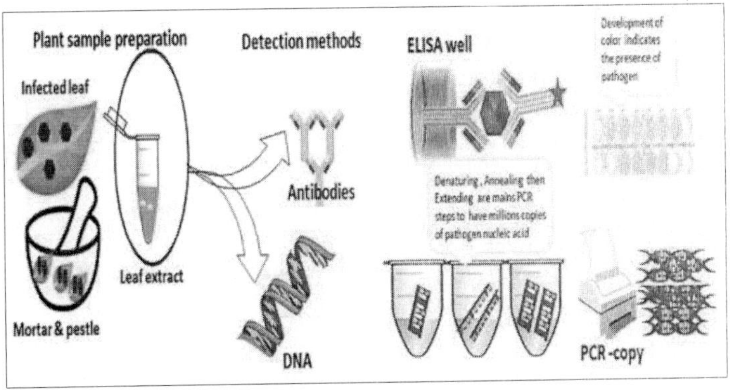

Fig 2: Schematic representation of the procedure for leaf extraction for pathogenic protein and DNA detection based on ELISA and PCR respectively (Scala et al. 2013).

Nucleic Acid Based Detection Methods

Nucleic acid based detection methods are mainly of two types i.e. DNA based (PCR and its variants like nested PCR (nPCR), cooperative PCR (Co-PCR), multiplex PCR (M-PCR), real-time PCR (RT-PCR), DNA fingerprinting and fluorescence in situ hybridization (FISH)) and RNA based (reverse transcriptase-PCR, nucleic acid sequence-based amplification (NASBA), and AmpliDet RNA). All of these methods can overcome uncertainty in disease diagnosis, aiding a rapid and accurate detection and quantification of pathogens. Nevertheless these methods are highly sensitive and specific, shortcomings are also exists.

In case of woody perennials, an uneven pathogen distribution can render molecular tools unreliable, particularly at pre-symptomatic stages. Pathogen titres are often below the sensitivity limit (usually <10 –100 pg) in seeds, insect vectors, water, and soil which may go undetected. False negatives can also occur when the DNA target sequence is degraded or reagents are of insufficient quality. Small sample sizes may falsify the real situation and sample cross-contamination can give false positives. Sensitivity limit of PCR-based methods is often lower than expected due to the presence of inhibitors of transcriptase and/or polymerases. PCR can amplify non-specific products due to mispriming or primer dimerization. In case of analysing the quarantine pathogens, sometimes dead pathogens can give false-positive results. Moreover, cost of the equipments and reagents also plays a vital role in selecting the desired detection methods.

Modern Methods of Plant Disease Detection

Traditional methods detect the pathogen only at later stages of the infection process when symptoms are visually seen. In contrast to that, modern/innovative methods allow detection of pathogen infection even the symptoms are unclear and present in only few plants. Also the modern methods are easy, rapid, inexpensive and reliable at presymptomatic to early developmental stages.

Modern day methods for plant disease detection includes, immunological and DNA-based methods (Lateral flow microarrays- LFM), methods based on the analysis of volatile compounds (VOCs) and genes as biomarkers of disease, remote sensing (RS) technologies in blend with spectroscopy-based methodologies, and sensors based on phage display and biophotonics (Martinelli et al. 2015). Lateral flow microarrays (LFM) are hybridization based nucleic acid detection method using easily visualized calorimetric signal. These are on par with microarrays in terms of detection limit and can reduce the expensive instruments cost. LFM relies on the availability of strong and reliable host and pathogen biomarkers discovered through transcriptomics or metabolomics. An integrated omics approach can detect the early infection of Huanglongping disease in citrus (Dandekar et al. 2010). Highly interactive proteins such as heat shock proteins or dehydrins, upregulated by different environmental factors (Natali et al. 2007), are possible indicators of plant health status.

Plants do emit volatile organic compounds (VCOs) in their surroundings which may play a vital role in their communication, growth, survival and defence (Baldwin et al. 2006). VOCs are low molecular weight biomolecules with a high vapour pressure and low boiling point. They freely exist in the gaseous phase under standard temperature and pressure and are generally present at ultra-low concentrations below the human olfactory threshold. VOC profiling is gaining important in the present scenario which helps in monitoring the health status of the plants under study.

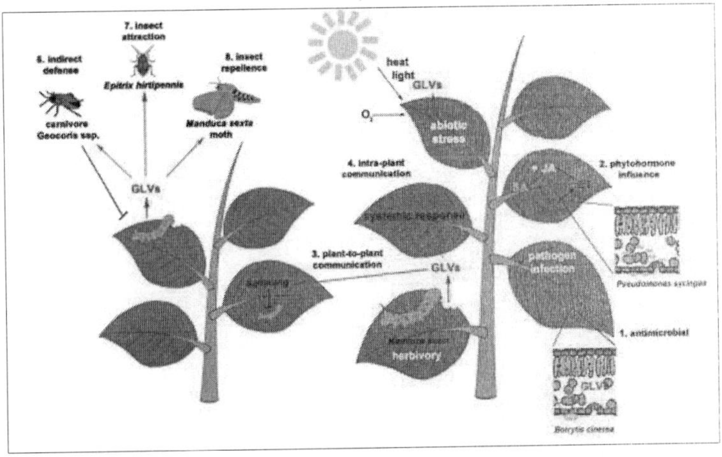

Fig 3: GLVs (Green Leaf Volatiles) are emitted during herbivory, pathogen infection and abiotic stress. GLVs have antimicrobial activity (1), influence phytohormonal networks (2) and are involved in plant-to-plant communication (3). GLVs can trigger a systemic defense response in un-attacked leaves of the same plant (4) and induce priming in neighboring plants (5). Some carnivores e.g. Geocoris ssp., prefer plants producing GLVs, thus representing an example of indirect defense (6). GLVs can be involved in attracting herbivorous insects,e.g., *Epitrix hirtipennis* (7) or repelling herbivorous insects, e.g. M *Manduca quinquemaculata* (8) (Scala et al. 2013).

A new avenue of research is opened by VOC profiling (Figure 3), which may detect mechanisms for "plant-to-plant" and "plant-to-insect" communication, gaining new insights into host responses to pathogens and abiotic stressors. VOC profiling is successfully applied in groundnut (*Arachis hypogea* L.). White mould (*Sclerotium rolfsii*) infected plants released some volatile compounds which may be preferred by beet armyworms which preferable consumes more infected leaves than non-infected. Similar studies were conducted in potato tubers inoculated with late blight (*Phytophthora infestans*) and dry rot (*Fusarium coeruleum*) after harvest identified 52 volatiles. This may helpful in post-harvest potato management against diseases (de Lacy Costello et al. 2001). VOC profiling study in Cucumber mosaic virus (CMV) showed insect vectors, *Aphis gossypii* and *Myzus persicae*, were preferentially attracted to CMV-infected plants. Remote sensing and spectroscopy-based technologies may speed diagnosis over traditional methods and enable better analysis of the spatial and temporal variability of disease, especially when calibrated and coupled with reliable reference information.

Methods under Development

Some of the innovative detection methods under development are described here.

1) Novel sensors based on the analysis of host responses, e.g., differential mobility spectrometer and lateral flow devices, deliver instantaneous results and can effectively detect early infections directly in the field;
2) Biosensors based on phage display and biophotonics can also detect instantaneously infections although they can be integrated with other systems; and
3) Remote sensing techniques coupled with spectroscopy-based methods allow high accuracy of results, these techniques may be very useful as a rapid preliminary identification of primary infections.

Whereas serological and nucleic acid based methods are the most accessible and effective to confirm disease diagnosis, volatile and biophotonic sensors deliver rapid results and may be used to find infections at asymptomatic stages. Remote sensing technologies will be enormously supportive to greatly spatialize diagnostic results. These innovative techniques signify unique tools to concentrate agriculture more sustainable and safe, avoiding expensive use of pesticides in crop protection.

Conclusion

The practical application of each traditional or innovative method depends on the cost and availability of instruments, spatialization level (plant, field, farm, or agricultural district), rapidity of analysis, and the stage of disease at which detection is possible.

References

1. Baldwin IT, Halitschke R, Paschold A, Von Dahl CC, Preston CA. Volatile signaling in plant-plant interactions: 'talking trees' in the genomics era. Science. 2006; 311:812-814. doi:10.1126/science.1118446

2. Camargo A, Smith JS. An image-processing based algorithm to automatically identify plant disease visual symptoms. Biosystem engineering. 2009; 102:9-21.

3. Scala A, Allmann S, Mirabella R, Haring MA, Schuurink RC. Green Leaf Volatiles: A Plant's Multifunctional Weapon against Herbivores and Pathogens Int. J Mol. Sci. 2013; 14(9):17781-17811.

4. Dandekar AM, Martinelli F, Davis CE, Bhushan A, Zhao W, Fiehn O *et al*. Analysis of early host responses for asymptomatic disease detection and management of specialty crops. Crit Rev Immunol. 2010; 30:277-289.

5. De Lacy Costello BPJ, Evans P, Ewen RJ, Gunson HE, Jones PRH, Ratcliffe NM *et al*. Gas chromatographymass spectrometry analyse of volatile compounds from potato tubers inoculated with Phytohthora infestans or Fusarium coeruleum. Plant Pathol. 2001; 50:489-496. doi:10.1046/j.1365-3059.2001.00594.x

6. Köhler G, Milstein C. Continuous cultures of fused cells secreting antibody of predefined specificity. Nature. 1975; 256:495-497. doi:10.1038/256495a0

7. Martinelli F, Scalenghe R, Davino S, Panno S, Scuderi G. Advanced methods of plant disease detection. A review. Agronomy for Sustainable Development, Springer Verlag/EDP Sciences/INRA. 2015; 35(1):1-25. <10.1007/s13593-014-0246-1.

8. Natali L, Giordani T, Lercari B, Maestrini P, Cozza R, Pangaro T *et al*. Light induces expression of a dehydrin-encoding gene during seedling de etiolation in sunflower (*Helianthus annuus* L.). J Plant Physiol. 2007; 164:263-273. doi:10.1016/j.jplph.2006.01.015